CANALS AND AMERICAN CITIES

ASSESSING THE IMPACT OF CANALS ON THE COURSE OF AMERICAN URBAN LIFE

Edited, with an introduction by Ronald C. Carlisle,
Senior Historian, Cultural Resources Section,
Michael Baker Jr., Inc.,
Pittsburgh, Pennsylvania

Papers from the Urban Archaeology Group's Symposium presented at the 24th Annual Meeting of the Society for Historical Archaeology and the 22nd Annual Meeting of the Advisory Council on Underwater Archaeology, Richmond, Virginia, January 9-13, 1991.

PUBLISHER

Canal History and Technology Press
Hugh Moore Historical Park and Museums, Inc.
P.O. Box 877, Easton, PA 18044-0877
(610) 250-6700

The cover is an artist's rendering of a dramatic moment in the daily life of a batteau crew that appeared in *Harper's Weekly* on February 21, 1874. More recently, the illustration has been used in promoting the yearly James River Batteau Festival, which was spawned by the archaeological discovery of sunken batteaux in the Great Basin of the James River and Kanawha Canal in Richmond, Virginia. This work is the subject of the article by William E. Trout, III and A. Howe Todd.

Easton, Pennsylvania, U.S.A.
ISBN 0-930973-14-3

CONTENTS

INTRODUCTION

Ronald C. Carlisle

From the late 18th century to the middle of the 19th, the United States, like France, England, Holland, and other European countries before it, was caught up in a brief but intensive experiment in canal building. Together with the construction of turnpikes, the digging of canals was the first truly national American phenomenon, a great if short-lived experiment in democracy and capitalism that attracted the attention and money of the most notable political and business leaders of the day. It was also the first great modification of the American landscape. An important motivation for the nearly simultaneous construction of canals in many states was the widespread belief in the intrinsic value and good of "internal improvements" as a way of stimulating both settlement and economic growth of the great American interior. Such improvements were "...widely regarded as essential to the preservation of republicanism" (Shaw 1990: 18). As many Atlantic seaboard states desired to participate in and, if possible, to prevail in the opening of the American West and its markets, competition among states also played a major role in the spread of the canal-building fervor. Individual entrepreneurs saw the opportunity to make money by reducing the cost of transporting food as well as timber, building stone, coal, and other raw materials from the American interior to the East Coast. The cost of transporting such products to market, not the cost of extraction, often proved the limiting factor in the development of business, and it was precisely this obstacle that the backers of early American canals hoped to overcome. Similarly, imported and American-made goods required a cost-effective means of distribution from the East Coast to the frontier. The limitations of overland travel were prohibitive in many cases given the technology of the day, and railroads, though soon to supplant canals as the distribution system of choice, were as yet in their technological infancy (see, among others, Rubin 1961; Goodrich 1961; Scheiber 1969; Porter 1973; Shaw 1990; Martin 1992).

The American canal-building era thus focused the debate for the first time over the respective roles of individuals and both the state and federal governments in promoting internal improvements and economic development. The concept of "public-private partnerships," a mainstay of contemporary American economic and civic life, was first articulated in the canal-building era as a means of paying the high cost of canal construction and maintenance. As a result of incurring these unprecedented costs, some states, of which Pennsylvania was the best example, soon also had to learn to cope with the demands of enormous public debt by levying income and property taxes. Important ethical and legal questions about collective good versus the rights of individual landowners were raised in "eminent domain" proceedings as the course of canals was plotted over the American landscape.

The canal-building period also stimulated the development of civil engineering, an all but non-existent American profession at the beginning of the 19th century. The construction skills first learned and subsequently shared among canal engineers as they moved from state to state and project to project soon created a small cadre of professionals who later applied their collective knowledge to the engineering of American railroad lines and to the laying out of new roads and cities.

In retrospect, many of the engineering accomplishments of the canal-building era now appear almost fantastic in view of the twin limitations imposed by the vastness and topographical barriers of the American interior and by an essentially preindustrial technology rooted in the use of natural construction materials—wood, stone, and clay.

The building of canals also had important consequences for American cities. Interurban competition for canal routes mirrored the competition among the states. As canals opened, prices for imported and native manufactured goods dropped and greater supplies of mineral wealth were channeled to cities, either as the ultimate consumers or as points of transshipment. For example, the anthracite coal of northeast Pennsylvania was rare in the homes and forges of nearby Philadelphia as late as 1815, but it was a commonplace in city homes just 15 years later when it was being shipped to that city by canal; moreover, canal-transported anthracite arrived in Philadelphia in sufficient quantities that it was economical to reload it onto sailing vessels for shipment to other coastal cities (Martin 1992: 139-140). Thus canals often stimulated urban economic growth both directly and indirectly.

Canals also promoted expansion of river traffic and trade. Shaw (1990: 157) has pointed out that before the opening of the Louisville and Portland Canal in December 1830, most Ohio River steamboats were limited to the stretch of the river below Louisville and the Falls of the Ohio. After 1830, however, 400 steamboats and 400 other craft passed through the canal to the upper reaches of the river, a number that increased to more than 1,000 steamboats per year by 1835, and that averaged 1,300 a year from then until 1851. This increase in Ohio River commerce had important consequences for both Cincinnati and Pittsburgh, one measure of which was the explosion in urban population that occurred during this period.

Canals were also significant in setting the stage for urban industrialization. The first Lake Superior iron ore was transported down the Erie Extension Canal to western Pennsylvania in 1853 (Shaw 1990: 78), and Great Lakes copper ore was brought by lake steamer to the Ohio and Erie Canal at Cleveland where it then found its way to Pittsburgh on the Ohio and Pennsylvania canal systems (Shaw 1990: 176). In the process, however, the opening of the Lake Superior copper mines also had a negative effect on the earlier copper industry in Maryland.

Canals and the canal-building era in American history have attracted the attention of historians and students of the history of technology for years. Perhaps the most reliable though least explored sources of information on American canals, however, are the canals themselves, many miles of which still dot the rural landscape or lie buried beneath urban rail lines. The archaeological excavation of canals thus offers another tool in the comparative study of these important vestiges of the American past.

The six articles published in this volume were first presented as symposium papers in Session 14 at the Society for Historical Archaeology's 24th annual meeting, which was held at the Marriott Hotel in Richmond, Virginia, on January 9-13, 1991. The session was scheduled for the morning and early afternoon of Friday, January 11, 1991. I organized and chaired the symposium on behalf of the society's Urban Archaeology Group. A particularly enjoyable and informative feature of the symposium was the bus tour of the James River and Kanawha Canal works in Richmond that took place between the morning and afternoon presentation of papers. The excursion was organized and led by William E. Trout, III, for the Virginia Canals and Navigations Society. The tour not only helped to illustrate Bill Trout's own symposium presentation on the Richmond Canal but also gave the symposium participants a first-hand opportunity to see how Richmond is utilizing its canal architecture to draw people to the urban waterfront once again.

Nor is Richmond alone in this new appreciation of canals and renewed interest in studying their role in defining the character of many 19th-century American cities. The comparatively recent archaeological rediscovery and excavation of urban transportation and waterpower canals, often the results of legally mandated cultural resources studies, have been accompanied in a few fortunate cases by the preservation, reconstruction, and public interpretation of these impressive masonry and wood constructions. Through this contribution to public history and to the promotion of tourism, canals once again can serve the economic interests of our cities. This aspect of canal research formed a dominant theme in the symposium that carries over into the final papers presented in this volume. Collectively, the articles present abundant evidence of the multiple roles that urban archaeologists are playing in excavating, researching, preserving, and adaptively reusing canal locks, aqueducts, bridges, and other canal features.

In retrospect, the title of the symposium, "Canals and American Cities: Assessing the Impact of Canals on the Course of American Urban Life," is too grand, but the articles in this volume do provide informative case studies of the archaeology of canals from Minneapolis, Pittsburgh, Alexandria, Richmond, and Charleston. In addition, Brian Morrell's article on the Delaware & Raritan Canal in New Jersey assesses the importance of that canal for the economic development of Philadelphia. Mark Newell's article on South Carolina's Santee Canal is one of the first archaeological reports on that early canal that was so important to the economic development of Charleston. Newell's paper is also an excellent example of the intellectual cross-fertilization that is possible between terrestrial and underwater archaeology. As originally planned, the symposium hoped to achieve even greater geographical diversity by including a paper by Michael S. Raber on the Enfield Canal in Connecticut and another paper by Edward Pershey of the Tsongas Industrial History Center on the waterpower canal at Lowell, Massachusetts. Unfortunately, due to contemporary economic hard times in New England, neither of these authors was able to participate, and our volume is diminished proportionately by their absence.

The discussants for the symposium included two leading experts on canals and canal technology, Lance E. Metz, historian at the Canal Museum in Easton, Pennsylvania, and Emory L. Kemp, recently president of the Society for Industrial Archaeology and now director of the Institute for the History of Technology and Industrial Archaeology in Morgantown, West Virginia. In reformulating the symposium papers for publication, I asked Lance and Emory, respectively, to compose a "Foreword" and "Conclusion" to the volume, based upon some of their comments in Richmond. I would especially like to thank them for their care in reading the original versions of the papers prior to the meeting and for their excellent discussions at the meeting itself.

References

Goodrich, Carter, editor

1961 *Canals and American Economic Development* by Carter Goodrich, Julius Rubin, H. Jerome Cranmer, and Harvey H. Segal. New York: Columbia University Press.

Martin, Albro

1992 *Railroads Triumphant: The Growth, Rejection & Rebirth of a Vital American Force.* New York: Oxford University Press.

Porter, Glenn

1973 *The Rise of Big Business 1860-1910.* Arlington Heights, Illinois: Harlan Davidson, Inc.

Rubin, Julius

1961 "Canal or Railroad?: Imitation and Innovation in the Response to the Erie Canal in Philadelphia, Baltimore, and Boston." *Transactions of the American Philosophical Society* 51(7). Philadelphia.

Scheiber, Harry N.

1969 *Ohio Canal Era: A Case Study of Government and the Economy, 1820-1861*. Athens, Ohio: The Ohio University Press.

Shaw, Ronald E.

1990 *Canals for a Nation: The Canal Era in the United States, 1790-1860*. Lexington, Kentucky: The University Press of Kentucky.

FOREWORD

Lance E. Metz

The effects that canals have had on America's national development have been largely overlooked by American historians. Canals are often treated as a footnote to railroads and river transportation, or are seen as a transitional transportation phase useful only as a harbinger of more modern and efficient forms. Yet such beliefs are not only erroneous but also self-defeating, since they retard scholarly efforts in the study of America's manmade waterways. Equally as important, such beliefs prevent historians from seeing canals in their true role as catalysts for change.

The construction of canals brought about the creation of the first commercially viable interregional transportation system for bulk commodities. However, even before this network was completed, canals had begun to make their impact on our national development. More than any other single factor, canals brought about the development of American civil engineering.

Beginning in the 1790s with the arrival of canal engineer William Weston, civil engineering skills and techniques first became known in this country through the construction of canals. It is remarkable that although Weston spent almost a decade in this nation, he was never able to complete any of the canals on which he served as chief engineer. Nevertheless, the skills he imparted to others, such as clay puddling, the proper use of a leveling transit, and the use of hydraulic cement, revolutionized the practice of canal building in the United States. Without Weston's input, it is doubtful that the Middlesex Canal, which linked the waterpower-rich Merrimack Valley with the seaport of Boston, and which also served as the inspiration for almost all subsequent North American canals, would have been completed. Thus, Weston's contributions to the development of American civil engineering cannot be underestimated, even though all of the projects that he directly supervised were either never completed or turned out to be commercial failures.

Weston's role as a transmitter of engineering techniques and construction technology was taken up during the early decades of the 19th century by Canvass White. This young American engineer brought back from a journey to England the secret of the production of hydraulic cement, as well as the most advanced canal construction techniques then available. White's imported knowledge was directly responsible for the completion of the Erie Canal, the commercial success of which launched a canal-building mania across this nation. Many of these waterways were engineered by White or others who had learned their skills under his direction.

Even railroads, which eventually replaced canals as the country's premier transportation system, were products of canal engineering, since it was William Strickland in his widely distributed and profusely illustrated report on English waterways that first brought railroads to the attention of North American entrepreneurs. Many early railroads in the United States were built as feeders to canal systems. These included the Lehigh Coal and Navigation Company's pioneering 1827 Summit Hill to Mauch Chunk line in Pennsylvania and the Delaware and Hudson Canal Company's line. In fact, the first steam locomotive to operate on a common carrier railroad made its experimental run in 1829 on the Delaware and Hudson Canal. John A. Roebling, whose name is most often associated with the construction of the Brooklyn Bridge, began the wire rope industry in the United States to supply this crucial commodity to the Allegheny Portage Railroad, which was a component of the Pennsylvania Main Line Canal. It was canal engineer John B. Jervis who made steam locomotive engineering possible in many parts of the United States through his invention of the turnable leading or "pony" track that enabled locomotives to traverse the badly built and often sharply curving tracks of early American railroads.

The catalytic role that canals played in the early national development of the United States is evident in the articles presented in this volume. Scott F. Anfinson, for example, highlights the tremendous impact that waterpower canals had on the industrial development of the Minneapolis area. By making waterpower plentiful and available, the waterpower canal system around the Falls of St. Anthony created the nation's most productive flour-milling district. Steven J. Shephard's article on the Alexandria Canal examines the interaction between that canal and its surrounding urban community. The articles by Shephard; Anfinson; Carlisle, Cowin, and Irion; and Trout and Todd also present interesting and informative discussions of the use of canals as urban historic resources that are frequently preserved and amenable to study, reconstruction, and interpretation within the context of American urban history. Mark Newell's study of South Carolina's Santee Canal provides important new information on an early chapter in America's canal-building history and its effect on the growth of one of its most important early cities, Charleston. Brian H. Morrell's article on the Delaware and Raritan Canal examines many of the pragmatic issues involved in the maintenance and restoration of an historic canal.

A common theme shared by all of the articles in this volume is the growing awareness of the many ways that canals and canal-related sites can be adaptively reused as recreational and historic sites. As these papers make clear, the sites are there—scattered throughout the eastern, southern, and midwestern parts of the country. What has been lacking in many areas has been public awareness of the important role that canals played in America's early national development. One may hope that the appearance of such recent publications as Ronald Shaw's *Canals for a Nation* and more specific case studies such as those in this volume will do much to address that lack of awareness and in the process increase both public and private resources that can be directed toward the preservation and historic interpretation of America's canals.

References

Shaw, Ronald E.

1990 *Canals for a Nation: The Canal Era in the United States, 1790-1860*. Lexington, Kentucky: The University of Kentucky Press.

MINNESOTA WATERPOWER CANAL
History, Archaeology, and Interpretive Potential

Scott F. Anfinson

Abstract

Minneapolis began as two cities on opposite sides of the Mississippi River. St. Anthony, on the east side, was founded first and began waterpower development first, but the west-side city of Minneapolis soon became the commercial and industrial center. The key to the success of Minneapolis was the early construction of a canal that allowed for efficient and intensive waterpower development. By 1880, Minneapolis led the nation in flour production largely because of the canal mills, which included the pioneer mills of Pillsbury and General Mills. After a century of intensive development, the mill district was largely abandoned in the mid-20th century. Redevelopment of the area in the 1980s has utilized archaeological investigations to examine the extensive canal and tailrace system as well as several mills. Revitalization of the area includes plans for an archaeological park with the canal at its heart.

Introduction

Two types of canals are discussed by the papers in this volume: transportation canals and waterpower canals. Although the lengths and immediate purposes of these canals are very different, they share several elements beyond the fact that they are artificial water channels. The ultimate aim of both types of canal is to improve efficiency for economic benefit. Economic benefits are both local and regional. Transportation canals initially benefitted the areas along their routes by providing construction employment, and, once completed, they stimulated the growth of cities as handling centers along the route. Through regional interconnections they ultimately led to widespread economic growth (Segal 1961).

Waterpower canals also provided local employment during their construction and then stimulated urban growth. Construction employment and urban growth were largely limited to a single city in the case of waterpower canals, but regional economic benefits were ultimately just as important as they were in the case of transportation canals. This is especially true with regard to Minneapolis, where the waterpowered sawmills processed lumber from the north and sent it to cities being built to the south. Waterpowered flour mills ground wheat from the West and sent it to the cities of the East.

Minneapolis began as two cities, one on each side of the Mississippi River at St. Anthony Falls. St. Anthony Falls, the only true waterfall on the Mississippi River, is about 30 feet high, and represents an impressive source of waterpower. The city of St. Anthony, on the east side of the river, had several initial advantages that should have established it as the namesake rather than its sister city of Minneapolis across the river. St. Anthony was settled first, it developed its waterpower first, and it was on the same side of the river as St. Paul, which was the head of navigation. Most of St. Anthony's first settlers were businessmen looking to establish a commercial center, whereas many initial west-side settlers were farmers. Yet, it was the west-side city of Minneapolis that gave the consolidated city its name and that was primarily responsible for the area's fame and fortune. The key to that fame and fortune was the west-side waterpower canal.

Over the last seven years, the Minnesota Historical Society (MHS) has been involved with an extensive program of urban archaeology along the central Minneapolis riverfront. This work began in response to a proposal to extend the West River Parkway through downtown Minneapolis. Because the project was to use federal funds, cultural resource review was required. The author of this paper was responsible for determining what impacts, if any, the West River Parkway project would have on archaeological sites.

At first glance the task seemed simple, especially to an archaeologist who had been trained at prehistoric sites on the Great Plains; an initial (but soon discarded) opinion was that the intense land disturbance in the heart of an urban center would certainly have destroyed most archaeological sites. It quickly became apparent that, from a conceptual point of view, Minneapolis, like other cities, could be thought of as a large, diverse, but rich and complex archaeological site. The central riverfront is perhaps the most complex and most archaeologically interesting area within this extensive city-site since it was the initial location of settlement and at one time was the city's industrial center.

In late 1982, the Minneapolis Park and Recreation Board (MPRB) selected a route for the West River Parkway that was the closest to the river of various alternatives (Figure 1). This route was chosen because it not only brought people close to the river, but excluded intervening development. It also had the greatest potential impact on archaeological sites, including the waterpower canal. A Memorandum of Agreement (MOA) was signed by the Park Board and the State Historic Preservation Office to insure that archaeological sites and historic structures would be protected. The MOA called for archaeological testing to find sites and for archaeological excavation of

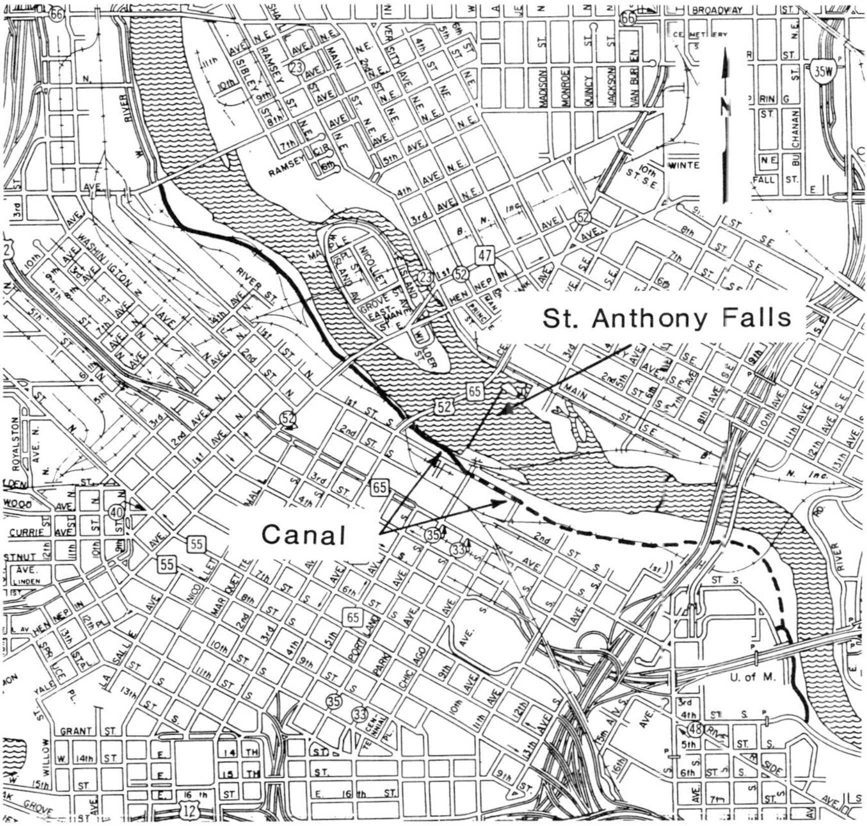

FIGURE 1. Map of the central Minneapolis riverfront showing the location of St. Anthony Falls, the west-side waterpower canal, and the route of the West River Parkway. The dark line shows where the Parkway was completed. The dashed line indicates where the Parkway was originally planned but has not yet been constructed.

important sites if damage to them by proposed developments could not be avoided. To their credit, the MPRB agreed to live up to the terms of the MOA even after federal funding for the project fell through prior to construction. The MPRB even proposed an "archaeological park" in the mill ruins lining the waterpower canal.

In order to properly assess the archaeological impacts of the route, the MPRB contracted with the MHS in the summer of 1983 to conduct archaeological testing within the proposed park boundary. The literature search documented that over 50 potential archaeological sites existed along the Parkway route (Anfinson 1984, 1989). Field investigations along the West River Parkway in 1983 located 21 sites even though no subsurface testing was done in the mill district where there is the densest site concentration (Tordoff 1984).

Archaeological testing continued in 1986 in conjunction with the start of construction on the north half of the Parkway project (Tordoff 1986; Tordoff and Clouse 1986). The 1986 excavations intensively examined several sites in the mill district and several sites that were unavoidable by the Parkway construction. Some construction monitoring was also carried out. The most recent archaeological work on the West River Parkway was done in 1989 (Szondy and Clouse 1990). This work focused on several flour mills along the waterpower canal in the proposed Mill Ruins Park. The northern half of the West River Parkway was completed in 1987. The southernmost segment was completed in 1990. Land acquisition for the intermediate segment through the southern mill district is still (1991) underway.

Historical Background

The construction of Fort Snelling eight miles south of St. Anthony Falls in the early 19th century established Euro-American civilization in Minnesota. In 1821, soldiers from Ft. Snelling erected a small frame sawmill on the west side of St. Anthony Falls. Two years later a stone gristmill was built adjacent to the sawmill.

The land east of the Mississippi River was ceded by the Dakota Indians in 1838. Franklin Steele, the sutler at Ft. Snelling, claimed the land directly east of St. Anthony Falls and began the private development of the area. Steele built a dam across the east channel just below Nicollet Island. At the east end of the dam he erected a small sawmill. Steele subdivided his land in 1849 and named his town St. Anthony. By 1850, 538 people lived in St. Anthony, and the town boasted a hotel and several businesses, including the sawmill. One year later a gristmill was built next to Steele's sawmill. The waterpower area of the east side was soon controlled by the St. Anthony Falls Waterpower Company, a partnership of Steele, several other St. Anthony businessmen, and several New York businessmen.

Following another treaty with the Dakota, the west side was opened to settlement in 1853. At first, the west side was envisioned as a farming community, but after the construction of a bridge over the Mississippi River in 1854, the land near the river soon became too valuable for farming. In the mid-1850s the original claimants subdivided their land into city lots. The especially valuable land nearest the Falls was consolidated into a partnership of 12 men who formed the Minneapolis Mill Company. The pace of settlement at St. Anthony Falls rapidly increased in the mid-19th century. On the east side, major building projects in the late 1850s included several flour and lumber mills. The 300 occupants of the west side in 1854 swelled to five times that number in two years. In 1856 the west-side city was incorporated under the name of Minneapolis.

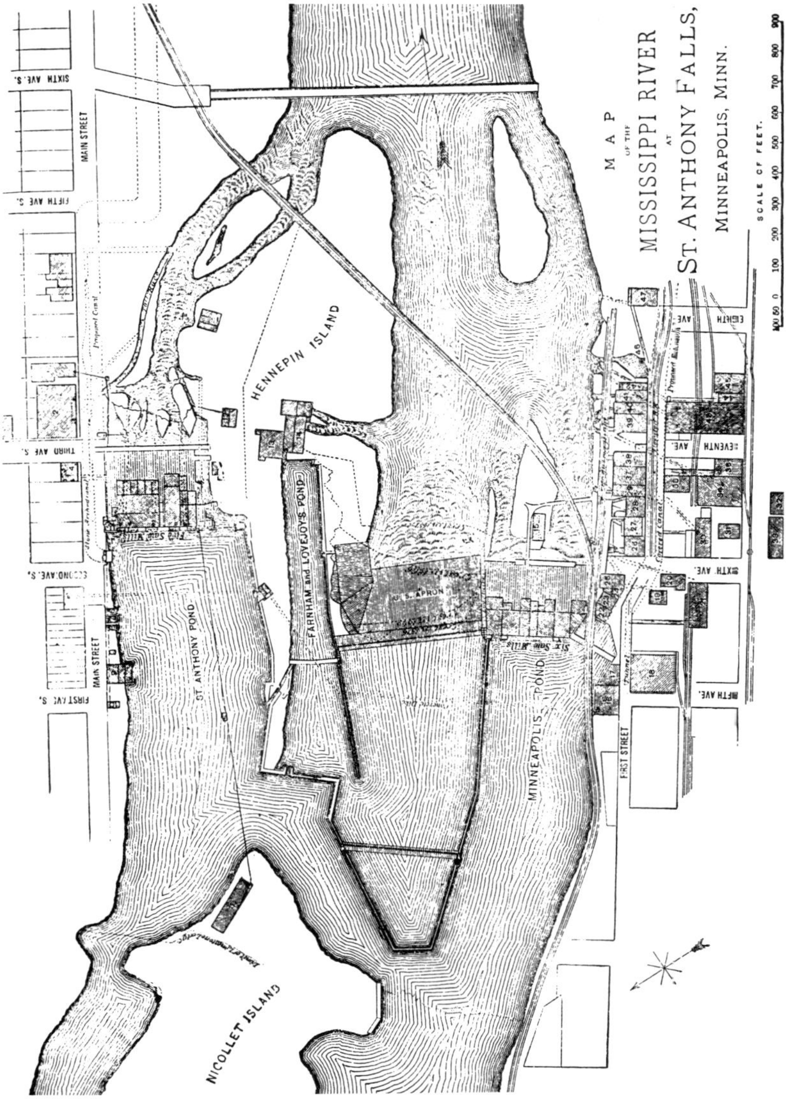

FIGURE 2. The dam complex and waterpowered industries at St. Anthony Falls in the early 1880s. The west-side waterpower canal is at the bottom center. *From Greenleaf 1887: 165.*

The developments that played the greatest roles in insuring the prosperity of the newly founded city occurred in 1857. A new dam was started on the east side of the river that extended out into the river's main channel from Steele's original dam. It was joined by a dam from the west side in 1858. The completed dam was V-shaped, pointing upriver, and channeled the water to mill ponds on both the east and west sides (Figure 2). In the 1860s, a wooden apron was placed over the waterfall to protect it from destruction. Although the dam has been replaced and repaired many times over the years, the basic configuration has remained essentially intact. The natural landmark of St. Anthony Falls has been replaced by the cultural landmark formed by the dam, apron, and spillways.

In addition to the dam, another major construction project was also begun in 1857. The Minneapolis Mill Company built a short canal that angled off the river for a block and then paralleled it for another block. In 1859 the first commercial flour mill was built along the canal. Development along the 400-foot canal was somewhat slowed by the Civil War, but by the end of 1865 a diversified west-side mill district was taking shape, and the canal had been extended downstream (south) another 600 ft. (Figure 3). Bordering it were six flour mills, a woolen mill, and an iron works. By the end of the 1860s, three more flour mills and a paper mill lined the canal.

On the east side of the river, in contrast, no new lumber mills and only two small flour mills were built in the 1860s. An attempt to expand the waterpower system on the east side failed. A tailrace tunnel beneath Main Street encountered a large natural cave and was abandoned in 1864. The only successful industrial development in the east-side mill district was the construction of three waterpowered iron works.

By 1870 the population of Minneapolis was over twice that of St. Anthony, 13,000 vs. 5,000. The mutual interest in developing and protecting the waterpower was a major factor in the consolidation of the two cities in 1872. The west side not only had more people, but had taken over the social and economic leadership of the area. The consolidated city was therefore called Minneapolis. Industrial development along the west-side riverfront was rapid in the 1870s. Fifteen new flour mills, a cotton mill, a new sawmill, the city waterworks, and an iron works were built adjacent to the canal. The flour mill development was spurred by the introduction of new technologies, many of them developed locally (e.g., the middlings purifier). Only three new flour mills were built on the east side in the 1870s.

By 1880 the population of Minneapolis was 47,000, and the city's flour had begun to gain an international reputation for quality and quantity. State-of-the-art flour milling could be seen at the new Washburn A and Crown Roller mills. Support industries, such as mill machinery works and barrel works, thrived. In 1880 Minneapolis led the nation in flour milling, a position that it held for the next 50 years.

Only a few new flour mills were built in the 1880s, notably the mammoth Pillsbury A Mill on the river's east side. A short canal was finally completed on the east side that supplied efficient waterpower to the Phoenix and Pillsbury A mills. The west-side canal was widened and deepened and the tailraces lowered to improve the waterpower. Continued growth of the flour industry soon depended on improvements in efficiency and milling

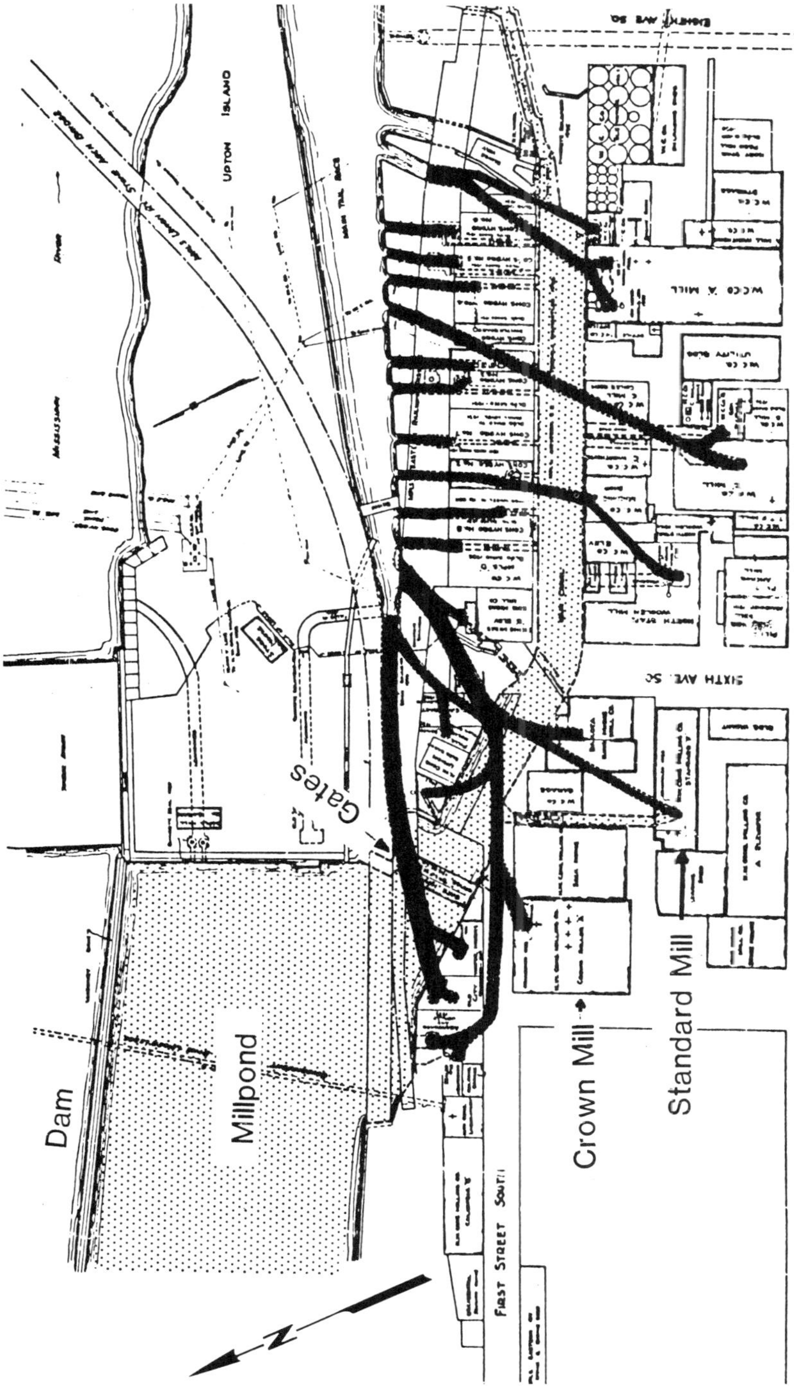

FIGURE 3. Diagram of the west-side mill district ca. 1926 showing the canal (stippled) and the tailraces (dark lines).

techniques as the waterpower was stretched to its limit. Rollers replaced the cumbersome mill stones, and electric lights made mill interiors safer. The waterpower was susceptible to low water, ice, and jamming caused by lumber refuse, so most of the mills added auxiliary steampower.

The last decade of the 19th century witnessed significant internal changes in the mill district. Ownership of most mills was consolidated into four large companies (two of which are now Pillsbury and General Mills). Perhaps the most significant change in the 1890s was the expansion of hydroelectric facilities. This changed the industrial landscape along the river, and the construction of a second dam expanded the waterpower district to the south.

In the early 20th century, Minneapolis led the nation in both flour and lumber production. By then, the sawmills were nearly gone from the St. Anthony Falls area. Most of them had moved upriver to steam-powered sites. Cylindrical concrete grain elevators (another local invention) became the new landmarks of the mill district. Annual flour production peaked in 1916 at over 18 million barrels. Soon thereafter, several of the smaller flour mills along the waterpower canal were torn down.

In 1930 Buffalo, New York, took over the national lead in flour production, and other mills along the Minneapolis waterpower canal were torn down. Minneapolis' population peaked at 521,700 in 1950, but its mill district was falling into disrepair. Only one flour mill remained on the river's east side and three on the west side in 1955. Construction of the St. Anthony Falls Lock and Dam complex was the death blow to the old mill district as it shut off waterpower to both sides of the river. The west side canal was filled in and its gatehouse torn down in 1960.

The last flour made on the west side came out of the Washburn A Mill in 1965. Today, only the Pillsbury A Mill still produces flour on the Minneapolis riverfront, and the mill is completely powered by electricity. Minneapolis still ranks third among American cities in flour production, largely due to several mills located off the river in the southeastern part of the city.

West Side Success, East Side Failure

A number of reasons explain why the west-side mill district of Minneapolis thrived while the east-side mill district of St. Anthony's struggled (cf. Kane 1987: 42-61). The late 1850s and early 1860s were difficult years due to the Panic of 1857 and the Civil War. Several of the owners of the east-side waterpower company were from New York and had little interest in the frontier town of St. Anthony. The west-side waterpower owners were mostly local men who were mill owners, and to them efficient waterpower and local growth were very important. In addition, the waterpower leasing system on the east side was inconsistent, whereas the west-side company copied its leasing system from the proven Holyoke, Massachusetts, system. The west-side company allowed only flour mill development near St. Anthony Falls and excluded industries that might be fire hazards. Commercial businesses were interspersed among the east-side waterpowered industries.

The greatest difference between the two sides of the river, however, was in the initial waterpower systems that each used. The east side was not actually a system at all but a hodgepodge of inefficient technologies that employed turbines located immediately adjacent to the Falls. Power was transferred through line shafts and overhead cables.

The west-side canal system was directly modeled after highly successful systems in use in the eastern United States at such cities as Holyoke, Lowell, and Lawrence, Massachusetts. The construction of the canal allowed intensive mill development to occur on either side of the canal. Within 15 years, the west side had 13 flour mills, 3 textile mills, an iron works, a sawmill, and a paper mill along the canal, while the east side had only 2 flour mills, 4 iron works, a sawmill, and a paper mill. Both sides had extensive platform sawmills built over the waterfall.

FIGURE 4. Looking north at the waterpower canal gates during their reconstruction in 1885. *Minnesota Historical Society.*

The west-side waterpower canal was continually improved from its initial construction in 1857 until its abandonment a century later. The system eventually consisted of a 2.9-mile maze of tunnels and canals. Water entered the main canal immediately above the dam. The flow was controlled by a massive masonry gate structure capped by a gatehouse (Figure 4). A small triangular area immediately downstream from the gatehouse was open, but after the canal intersected 1st Street, it was covered with a wooden deck. This deck extended for two blocks downstream along 1st Street parallel to the river; parts of the deck were eventually replaced with reinforced concrete. Above the deck was a railroad trestle.

FIGURE 5. Looking southeast at the waterpower canal during its reconstruction in 1885. Note the headrace openings along the canal wall and the railroad trestle being built above the canal. *Minnesota Historical Society.*

Along the canal near each mill, a headgate let water into vertically set turbine tubes (Figure 5). Mills not immediately on the canal were connected to the main canal by smaller spur canals or tunnels. Once inside the mills, the water fell down the draft tubes, turning the turbines, and then entered the tailrace system 40-50 feet below street level. The tailrace system brought the water back to the river below the Falls.

Within 15 years of the canal's construction, the two cities located at St. Anthony Falls were consolidated. That the new city was named Minneapolis, not St. Anthony, is a testimony to the success of the waterpower canal. By the time the east side finally constructed a short headrace tunnel in 1880, much of the land near the Falls was no longer available for mill development; only two flour mills ever used the east-side tunnel for power.

Archaeological Investigations

Recent archaeology in the west-side mill district has focused on several mills, the canal gates, and the tailraces (Tordoff and Clouse 1987; Szondy and Clouse 1990). Unofficial construction monitoring on various private and public developments has documented other mill and canal features (Anfinson 1990). Five examples of the archaeological investigations are discussed below.

Canal Gates

The initially preferred West River Parkway route intersected the historically documented location of the waterpower canal gates; thus, excavations were undertaken in 1986 to confirm the location of the gates and to assess their condition. At that time, the gates lay beneath a restaurant parking lot. Once the gates were exposed and shown to be in good condition, the alignment of the parkway was shifted to miss them.

The 1986 excavations revealed portions of four of the eight gate openings. Horizontal exposure of the gates was limited by modern surface features. Excavations extended to a depth of 20 feet, but the bottom of the gate structure was not uncovered due to the limited reach of the backhoe and the high watertable. The archaeological work showed that the gateways were arched and built of granite. Four courses of sandstone laid immediately above the arches directly supported the iron gate frames. Double limestone walls topped the structure, and a ballast chamber between them was filled with limestone rubble. The iron frames on the north side once guided wooden gates (which were not preserved) up and down to control the water flow.

Tailraces

Southeast of the canal gates, an excavation unit exposed a portion of the circa 1890 tailrace outlets that had been sealed and then filled over during the Upper Lock construction in the early 1960s. The fine sand that covered the area made backhoe excavation very difficult, and the walls continually collapsed. Thus only a small portion of the tailrace could be exposed. The top of the mortared limestone tailrace wall was revealed 18 feet below the modern ground surface. The lower 3-4 feet of the wall had a concrete facing. The bottom of the tailrace was covered with water, but the raceway appeared to be about 8 feet deep.

The northern interior of the tailrace system was also explored and is essentially intact, although much of it has been flooded by storm drains and sanitary sewer breaks. At least half of the system is flooded to a depth of 8 feet. Portions of the system could be reached only by rubber raft. The problem of access to the tailraces and transportation problems within the system made detailed surveying and photography very difficult. The tailrace tunnel system has also served as a utility access to sewer lines, water pipes, and natural gas pipes carried along the tunnel ceiling.

The tailrace tunnels were excavated into soft sandstone using the overlying limestone as a natural ceiling. The walls were first lined with masonry and later with cement to prevent erosion. The oldest tunnels, dating to 1859, are lined with worked limestone set in a good-quality mortar. Rounded alcoves in the tunnels contain 4-foot-diameter iron draft tubes that once led to the mills above. The mouth of the oldest tunnel is formed by a yellow brick arch. The exit has been sealed by cinder blocks.

The main tunnel at the north end of the tailrace system runs directly beneath the main waterpower canal. The walls of the tunnel are made in sections of limestone blocks and formed ferroconcrete. Stonework dominates the lower part of the tunnel. The wooden catwalk that once ran through the tunnel has been reduced to a jumble of rotting timbers. Numerous forks are present along this tunnel, some lined with limestone masonry, concrete, and even glazed brick. Concrete or brick pillars supporting the natural limestone ceiling are occasionally found in the wider sections of the tunnel. At the upper end of one of the branch tunnels are three draft-tube alcoves lined with glazed brick. Unlike other alcoves, there has been little silting here, and the bottoms of the draft tubes are not buried. The tubes stop about 2-3 feet short of the floor, allowing for water outflow. A deflection cone is set in a curved basin in the floor beneath the tubes to reduce erosion.

Canal Deck

Engineers for the West River Parkway segment through the mill district were concerned that the concrete deck over the northern part of the waterpower canal was not structurally sound. The concrete deck had replaced a wooden deck in 1934. In October of 1986, demolition of the deck exposed the sand-filled canal. The deck was much sounder than predicted and proved very difficult to remove. It was supported by large reinforced concrete beams interconnected with concrete-covered steel I-beams.

Standard Mill Canal Spur

Construction in 1987 of a plaza for a hotel being built in the abandoned Standard Flour Mill exposed the headrace canal to the mill. Prior to the construction, most of this spur canal was covered with a concrete deck, although part of the canal was visible through a steel grate. The plaza construction removed the deck and revealed that the top of the headrace had concrete beams similar to the deck support for the main canal. It was also discovered that a trash gate had been placed in the raceway to prevent debris from entering the turbines.

Crown Mill Headrace

Sewer construction along the street over the canal in 1987 destroyed a portion of the granite-paved street surface and cut through a large portion of the Crown Mill headrace. Unlike the Standard Mill headrace, the Crown Mill headrace was a completely covered, brick-faced arched tunnel, the top of which was about 6 feet below ground surface.

Riverfront Renewal and Interpretive Potentials

The industrial death of Minneapolis' central riverfront in the mid-20th century left the heart of the city a wasteland. Abandoned buildings and vacant lots were avoided by citizens and visitors alike. But with the abandonment also came opportunity. The central riverfront was, for the first time since the 1850s, open for non-industrial development.

The recolonization of the west-side central riverfront began in 1968 with the opening of a restaurant built into some mill ruins. Across the river, another restaurant opened in 1973 in a building that had been built in the 1890s as a saloon. Four years later, a shopping center opened in an old mattress factory. Recreational opportunities were also being developed. A small park was established just southeast of the mill district in 1971. The redevelopment of the central riverfront began in earnest in the 1980s. Housing became available with the opening of several hotels and apartment buildings. Office space expanded into new buildings and converted mills. Three of the four central riverfront highway bridges were replaced. Parkland expanded along both sides of the river.

Awareness of the historical importance of Minneapolis' central riverfront was highlighted in 1971 when the St. Anthony Falls Historic District was entered on the National Register of Historic Places. One year later, the Minneapolis Heritage Preservation Commission was established. While these acts provided some protection for riverfront historical resources, before the West River Parkway archaeological excavations historic preservation focused exclusively on standing structures. Furthermore, little had been done to promote the historic interpretation of the area. As of 1990, on-site interpretation of riverfront history was still largely limited to captions on photographs that hang on restaurant walls.

One of the difficulties in promoting historic preservation and interpretation along the riverfront has been the multitude of agencies and companies involved with riverfront management. The City of Minneapolis, various local, state and federal agencies, and private concerns all were interested in the development of the riverfront, but all have their own agenda. Lack of coordination was the principal roadblock preventing substantive and comprehensive historic interpretation of the area.

The state and local agencies finally got together and formed the St. Anthony Falls Heritage Board in 1988. This board has produced an interpretive framework (Ellerbe Becket, Inc. 1990) and is moving ahead with a heritage trail system and plans for an interpretive center. Public and private cooperation have been stressed. The premise is that successful development of the central riverfront can only come about through a mix of commercial development, recreational opportunities, and historical interpretation.

A key element in the historic interpretation of the Minneapolis riverfront is its archaeological resources. The West River Parkway excavations demonstrated that important, well-preserved archaeological features were indeed preserved along the central riverfront. In the mill district, these resources include a two-block stretch of mill foundations along the east side of the waterpower canal in addition to the canal, its control gates, and tailraces. Most of these features will ultimately be exposed, creating an extensive archaeological park where the public will be able to better appreciate and understand the industrial processes and waterpower systems. Most of all, it will allow visitors to once again appreciate the scope of industrial development that established Minneapolis as the “Mill City” and the site of perhaps the greatest direct-drive waterpower center the world has ever known.

Conclusions

Geographically well positioned between natural resources and markets, St. Anthony Falls rapidly became more than a “landmark in the wilderness.” In the mid-19th century Minneapolis had developed into a waterpowered manufacturing center. The early construction of the west-side canal was responsible for the intensity and success of the milling industries at the Falls. By 1880, tall flour mills lined the canal and brought the city international fame and fortune. The industrial heyday at the Falls lasted just over a century before the central Minneapolis riverfront was abandoned.

Redevelopment of the central riverfront in the 1970s and 1980s enjoyed only mixed success and in the process destroyed many historical resources, especially archaeological sites. The industrial character of the area was compromised by superficial attempts to make it "attractive." Construction often proceeded with little thought of practical needs, much less riverfront history. A high rate of office and apartment vacancies demonstrated that commercial development alone was not the solution to riverfront revitalization.

Under the leadership of the Minneapolis Park and Recreation Board, parkland has now reappeared along the riverfront, recreational opportunities have flourished, and archaeological potentials have been demonstrated. City officials and most private developers are now much more sensitive to the historical resources of the area, not only for the sake of preserving elements of the city's past, but to further the commercial attractiveness and recreational possibilities in the contemporary city.

The stage is now set for historical interpretation, an act long overdue in the Minneapolis riverfront drama. The St. Anthony Falls area is arguably the most important historic site in Minnesota. The waterpower canal is at the heart of that history and at the heart of the interpretive plan for the central riverfront. As it did in the 19th century, the canal is once again fostering economic prosperity but in a manner never anticipated by its builders. Rather than the flow of water through its granite gates, it is now the flow of people around the canal that will help make the riverfront once more successful.

References

Anfinson, S.

1984 *Archaeological Potentials on the West Side of the Central Minneapolis Waterfront.* St. Paul: Minnesota Historical Society.

1989 *Archaeology of the Central Minneapolis Riverfront, Part 1: Historical Overview and Archaeological Potentials.* The Minnesota Archaeologist 48 (1-2).

1990 *Archaeology of the Central Minneapolis Riverfront, Part 2: Archaeological Explorations and Interpretive Potentials.* The Minnesota Archaeologist 49 (1-2).

Ellerbe Becket, Inc.

1990 *St. Anthony Falls Interpretive Plan.* Minneapolis: St. Anthony Falls Heritage Board.

Greenleaf, J.L.

1887 "Report on the Water-Power of the Mississippi River and Some of its Tributaries." *Tenth Census of the United States, 1880*, Volume 17: 119-276. Washington, D.C.: U.S. Government Printing Office.

Kane, L.

1987 *The Falls of St. Anthony: The Waterfall that Built Minneapolis*. St. Paul: Minnesota Historical Society.

Segal, H.

1961 "Canals and Economic Development." In *Canals and American Economic Development*, edited by C. Goodrich. New York: Columbia University Press, 216-247.

Szondy, D., and R. Clouse

1990 *Archaeology Along the West River Parkway, 1989, Minneapolis, Minnesota*. St. Paul: Minnesota Historical Society.

Tordoff, J.

1984 *A Phase 1 Archaeological Survey of the West River Parkway, Minneapolis, Minnesota*. St. Paul: Minnesota Historical Society.

1986 *Phase 1 Archaeological Testing of the Fuji-Ya Parking Lot and Palisade Mill Sites*. St. Paul: Minnesota Historical Society.

Tordoff, J., and R. Clouse

1986 *Archaeological Excavations Along the Proposed West River Parkway 1986, Minneapolis, Hennepin County, Minnesota*. St. Paul: Minnesota Historical Society.

PENNSYLVANIA CANAL, PITTSBURGH
Excavation, Preservation, and Interpretation

Ronald C. Carlisle
Verna L. Cowin
Jack B. Irion

Abstract

Substantial portions of lift lock number 4 and an adjacent weigh lock on the Western Division of the Pennsylvania Canal were excavated in 1987 during road construction in Pittsburgh. The discovery of these archaeological features established an important local preservation challenge and opportunity, not only to record the canal features but to carefully dismantle, remove, and conserve selected architectural elements for eventual reconstruction in an urban canal park. Preserving these large-scale elements of material culture is difficult to coordinate and to finance, but it is vital for interpreting the canal's significance in Pennsylvania history and for understanding the canal's contribution in Pittsburgh's economic and social growth.

Introduction

Pennsylvania's earliest experiment with canal building was the 1797 construction of a canal around Conewago Falls on eastern Pennsylvania's Susquehanna River near the site of Three Mile Island (Burkett and Ilisevich 1983). Despite this early beginning, it was the spectacular success of New York's Erie Canal, opened in 1825, that drew national attention to the possible economic benefits of slackwater navigation and launched a flurry of building activity in surrounding states. The Erie Canal became a popular trade route for imported and Atlantic Coast goods that were increasingly in demand by a westward-moving American frontier.

Merchants in Pennsylvania and Ohio, for example, soon pressured their state legislators to build canals in their respective states in an attempt to direct passengers and freight through Pittsburgh, Youngstown, Cleveland, and other emerging towns and cities.

The Pennsylvania Main Line Canal was authorized by three legislative acts between 1824 and 1827. The canal eventually crossed nearly all of Pennsylvania and consisted of three major sections, the Eastern, Juniata, and Western divisions (Figure 1). In addition, the Columbia and Philadelphia Railroad and the Allegheny Portage Railroad provided important links

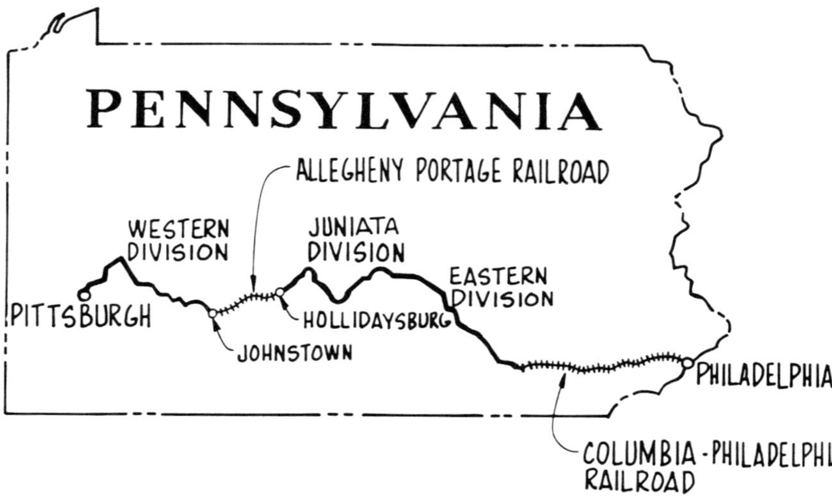

FIGURE 1. Map showing the three main divisions of the Pennsylvania Main Line Canal as well as the Allegheny Portage Railroad and the Columbia-Philadelphia Railroad. *Historic American Engineering Record, National Park Service. Delineated by Paul R. Pinchot, 1988.*

to the canal segments. The Allegheny Portage Railroad, for example, connected the Juniata and Western divisions of the canal. This railroad link was completed in 1834 and allowed canal boats to be lifted across the Allegheny Mountains from Hollidaysburg to Johnstown, a distance of 26 miles (Rhoads 1960: 8). From Johnstown (which later in the 19th century became a major iron and steel center), westward-moving canal boats could proceed through the Conemaugh, Kiskiminetas, and Allegheny river valleys to Pittsburgh.

Although the Pennsylvania Canal terminated in Pittsburgh and the adjacent town of Allegheny (today's Northside neighborhood of Pittsburgh), canal passengers and freight could continue west on Ohio's canal system, which took shape at the same time that canal building was underway in Pennsylvania. The members of Pittsburgh's business community were anxious to see Ohio construct canals in the eastern end of that state. They recognized that Ohio canals would form important westward extensions of Pennsylvania's own canal system and would help to market Pennsylvania, particularly Pittsburgh, products (Reid 1983: 6; Scheiber 1969: 103). The Pennsylvania and Ohio Canal (also known as the Mahoning or "Cross Cut" Canal), the Sandy and Beaver Canal, and the Hocking Canal were therefore built to provide links between the Pennsylvania Main Line Canal and the Ohio and Erie Canal. Members of Philadelphia's wealthy mercantile houses also played an important role in an 1833 convention that debated the merits of various routes for these canals. Thus, the influence of the Philadelphia and Pittsburgh urban and industrial élite on the decision to build canals and where to build them was apparent from the onset of the canal-building era in the mid-1820s and extended beyond the political borders of Pennsylvania itself.

This outward-looking and expansionist mentality can be said to be the initial and perhaps single most important contribution that the canal era made to the synergy of 19th-century urban life. The local economic importance of the Pennsylvania and Ohio Canal, to discuss just one of the examples by which Pittsburgh was linked to the West, was tremendous. The route through the Mahoning Valley offered Cleveland and interior Ohio merchants access to Pittsburgh markets and extended from there down the Ohio and Mississippi rivers to New Orleans. The Pennsylvania and Ohio Canal also played a key role in the industrial and economic development of Youngstown, Ohio, as the canal linked that city with the increasingly important Brier Hill, Ohio, coal mines, which in 1840 were producing about 700,000 bushels of coal per year. Due to Cleveland's demand for coal as both a domestic and naval fuel and to the presence of the canal, the output of the Brier Hill mines increased to 2,000,000 bushels per year by 1848 (Scheiber 1969: 215-216; Carlisle and Dalesandry 1988: 2.15-2.16).

Construction of the Pennsylvania and Ohio Canal also stimulated the production and transportation of cheese and other dairy products as well as wool and agricultural produce. These products were soon being shipped north to Cleveland, east to Pittsburgh, northeast to New York, and south to Cincinnati. Pennsylvania products also moved into Ohio along this canal. In 1840, for example, 4,000,000 pounds of Pittsburgh ironwares were transported by slackwater canal. Hardware, farm implements, and other iron goods amounting to 12,000,000 pounds were carried just five years later (Scheiber 1969: 216-217).

Many other examples of the importance that the relatively brief canal era had on the development of Pennsylvania cities and towns could be given. The Delaware and Hudson Canal, for example, played a large role in the economic growth of the Lackawanna River Valley, and northeastern Pennsylvania in general, by opening up the rich anthracite coal veins of that area to east-coast markets. The eastern Pennsylvania coal towns of Carbondale, Honesdale, and Hawley owe their establishment or growth to the completion of this canal in 1828 (Humphrey 1990).

Dozens of other Pennsylvania towns were boosted by the arrival of the Pennsylvania Main Line Canal. Saltsburg, on the north side of the Kiskiminetas and Conemaugh rivers in Indiana County, Pennsylvania, and Alexandria, on the Juniata River in Huntingdon County, are two good cases in point. The impact of the arrival of the canal on commerce, population, and architecture has recently been assessed by a National Park Service (HABS/HAER) study

(Leach 1989). Saltsburg was established before the canal, but transportation of its major commodity, salt, was significantly improved by this engineering accomplishment. Here, as in many Pennsylvania canal towns, the effect of the canal went beyond improved transportation for agricultural, industrial, and mineral goods. The location of the canal parallel to the rivers essentially usurped the earlier importance of the rivers as the geographical locus of both residences and commercial ventures. As it was also close to the river, the canal benefitted from proximity to the many water-powered mills located there or on nearby tributary streams. A reciprocal benefit accrued to the mill owners, tanners, and other businessmen who relied upon access to water in the production process: by having the canal close at hand transportation costs from their places of business to the canal were minimized.

Warehouses were needed to store products until they could be shipped on the canal, and these buildings therefore became important landscape elements in the near-canal areas of many Pennsylvania towns. Competition for scarce real estate near canal basins and warehouses increased the price of urban land and buildings in these areas of town, and stimulated speculation.

The bustling dockside area of the canals and canal basins sparked a host of support and service businesses, including hotels, inns, cooperages, rope-making businesses, etc. Boatbuilders, carpenters, furniture makers, and other woodworkers found both a ready market for their products and a pool of skilled wage labor in the near-canal neighborhoods.

The thousands of (largely Irish) unskilled and semi-skilled immigrants who dug the Pennsylvania canal by hand were themselves attracted to the towns and cities through which it passed. Their numbers swelled the population of these towns, where they helped to form the demographic base of the urban proletariat in the next generation.

The canal-building era in Pennsylvania therefore had both direct and indirect consequences on the course of urban life that went beyond the economic viability of the canal itself. Compared to the Erie Canal, for instance, the Pennsylvania Canal was never an economic success. Nevertheless, its construction produced the first major transformation of the physical and cultural landscape of Pennsylvania, a process that affected literally the entire length of the Commonwealth. The lessons learned in financing, building, and operating the canal stimulated an outward-looking, largely urban point of view that helped to set the stage for the massive extractive and manufacturing industries of the second half of the 19th century.

Building the canal also demanded sophisticated civil and hydraulic engineering skills and produced a market for those with the requisite formal education to construct and maintain it. Managers with the ability to coordinate the activities of private contractors and the thousands of skilled and unskilled laborers who were needed to dig the canal prism, build the locks, and erect the support structures that made the canal functional were also in demand.

Choices about the routing of the canal to one town at the expense of another were perceived as decisions that could have important economic consequences, as the dispute between Pittsburgh and Allegheny for the location of the western terminus of the Pennsylvania Canal attests. Thus, the canal-building period also stimulated inter-urban competition. Once built, the canal promoted the establishment or expansion of diverse urban waterfront businesses and support services geared to the transportation of both products and passengers. The increase in passenger travel expanded inter-urban sharing of ideas and values, a process that not only produced urban sophistication but which further accentuated the differences between town and rural dweller. The demise of the canal, when it came, often meant an economic downturn for these towns and cities, especially if the canal's direct and indirect benefits were not assumed by the railroad.

In the remainder of this article, the rediscovery, excavation, study, conservation, proposed reconstruction, and interpretation of two locks from the Western Division of the Pennsylvania Canal in Pittsburgh are discussed. In so doing, the potential for canals to contribute once again to urban life, this time as important cultural resources, is also discussed.

The Western Division of the Pennsylvania Canal

Construction of the Western Division of the Pennsylvania Main Line Canal was authorized by the state legislature in February, 1826. The section was built at a cost of nearly $4,000,000, and the first canal boat, the *General Lacock*, reached Allegheny Town (today's Northside area of Pittsburgh) in July, 1829. By 1834, through-traffic to Philadelphia had been established. When completed, the Western Division was 104 miles in length; it had 11 dams, 65 locks, 43 culverts, 46 overfalls, 16 aqueducts, 18 towpath bridges, 89 road and farm-lane bridges, and two tunnels. At Pittsburgh, lift locks lowered the canal approximately 39 feet to the level of the Allegheny and Monongahela rivers.

The canal approached Pittsburgh on the northern side of the Allegheny River but split a short distance above the confluence of the Allegheny and Monongahela rivers (Figure 2). An aqueduct (which collapsed in 1861 and was never repaired) crossed over the Allegheny River and gave Pittsburgh its own canal outlet complete with a series of canal basins. On the Pittsburgh side of the river, the canal was then carried in a tunnel dug through Grant's Hill; it emerged on the Monongahela River through a series of four locks near Suke's Run. On the Allegheny Town side of the Allegheny River, four

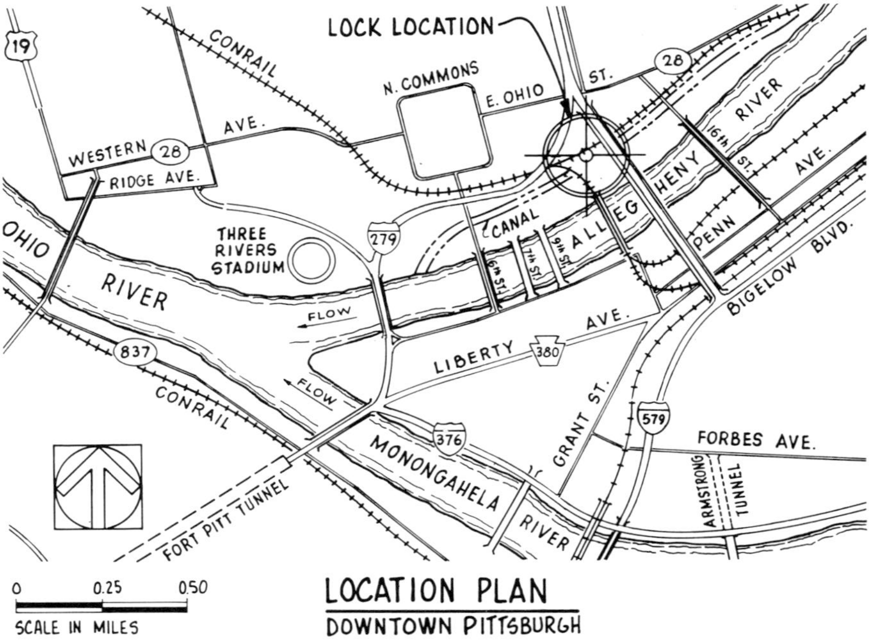

FIGURE 2. Location of the canal excavations in Pittsburgh in 1987. When the canal was built, these locks were in the town of Allegheny. Today, this is the Northside neighborhood of the City of Pittsburgh. *Historic American Engineering Record, National Park Service. Delineated by Paul R. Pinchot, 1988.*

other locks led from the area of the aqueduct to canal basins near Cremo Street (McCullough and Leuba 1962; Burkett and Ilisevich 1983; Shank 1984a, 1984b; Ilisevich and Burkett 1985).

As was true of many other canals, the Western Division of the Pennsylvania Canal soon yielded to competition from the railroad. The Pennsylvania Railroad acquired the canal right-of-way for $7,500,000 in 1857, and between 1861 and 1865, the canal in Pittsburgh ceased operation. The railroad often filled in the canal prism and laid its tracks over it, ironically insuring the preservation of many canal features, such as the two locks uncovered in Pittsburgh in the summer of 1987 during construction of the I-279/579 project (the East Street Valley Expressway) by the Pennsylvania Department of Transportation with funding from the Federal Highway Administration.

The location of the two locks, which included lift lock number 4 and its attendant weigh lock (Figures 3, 4), had been known for some years. The Carnegie Museum of Natural History, for example, had conducted a Phase I archaeological survey along portions of the proposed route of the East Street Valley Expressway in 1980 and had identified locations for the canal route and related structures on the Northside (Lantz 1980). Carl K.

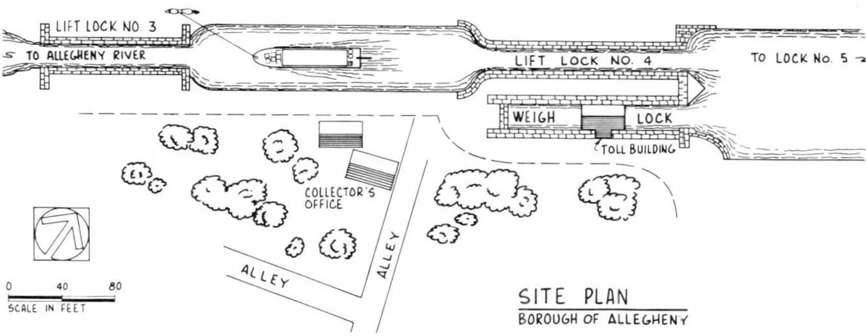

FIGURE 3. Site plan of lift lock number 4 and the adjacent weigh lock. Part of the foundation of the toll collector's house was also identified archaeologically. Lift lock number 3 was not excavated in 1987 but may be intact beneath the railroad right-of-way. *Historic American Engineering Record, National Park Service. Delineated by Paul R. Pinchot, 1988.*

Burkett, Jr. and Robert D. Ilisevich subsequently completed an exhaustive survey of the historical sources relating to the Pennsylvania Main Line Canal in Pittsburgh in 1983 in anticipation of a number of bridge, highway, and redevelopment projects (Burkett and Ilisevich 1983).

Archaeological testing of the Interstate 279/579 project right-of-way in 1986 encountered foundations and other structures, but the fact that any canal features preserved lay beneath nearly 20 feet of fill did not permit opening of sufficient ground to verify their presence. At this point, a decision was made to postpone the archaeology until certain alterations, including the relocation of a portion of a railroad bed, exposed a large enough area to provide access to any remaining canal features.

In the summer of 1987, the lift and weigh locks were uncovered, almost intact, within the PennDOT right-of-way by archaeologists from GAI Consultants, Inc., the archaeological contractor to PennDOT on the project (Irion 1989). The project also included a Level 1 Historic American Engineering Record documentation of the locks (Figures 5, 6).

In terms of the technology used in their construction, the two locks present very different pictures. The lift lock is fairly typical. It consisted of a chamber 90 feet long, about 14 feet deep, and 15 feet wide, 2 feet narrower than the locks found in the Eastern Division of the canal (see Figures 3, 4).

Construction of the lift lock began by excavating the lock pit. Foundation timbers measuring 34 feet in length and 1 foot square were then placed across the width of the lock pit at no more than 1-foot intervals. Gray puddling clay was then packed between the timbers (see Figure 6), and a floor of 2-inch-thick pine boards was spiked to the foundation timbers. The stone chamber of the lock was then built on this wooden foundation and consisted of two parallel stone walls 138 feet long, 15 feet apart, with

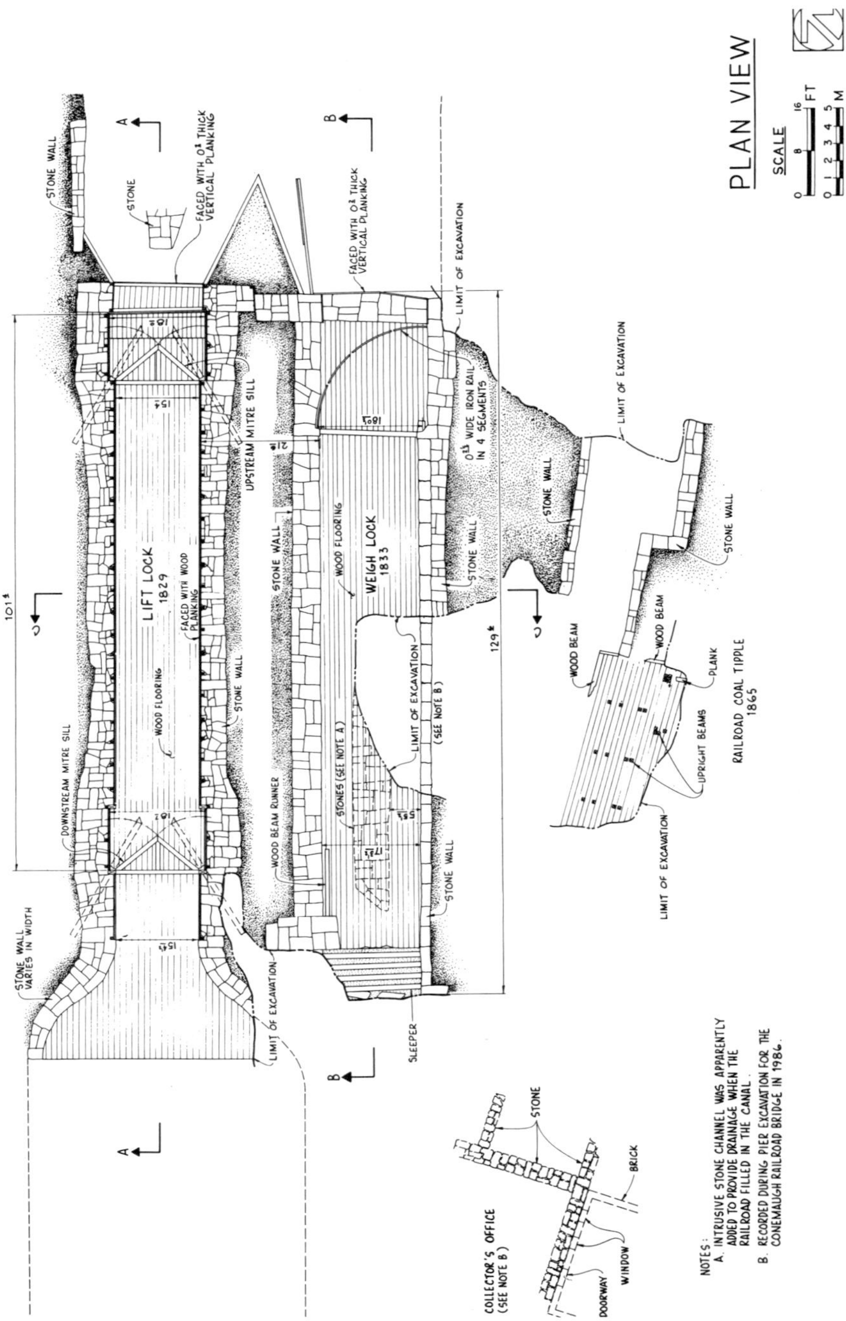

FIGURE 4. Detailed plan view of lift lock number 4 and the adjacent weigh lock based upon the archaeological excavations in 1987. *Historic American Engineering Record, National Park Service. Delineated by Paul R. Pinchot, 1988.*

FIGURE 5. Downstream sandstone wing wall of lock number 4 looking east into the lock chamber. After archaeological excavation, the sandstone was numbered and removed off-site for eventual reconstruction of an urban canal park. *Photo: Ronald C. Carlisle.*

FIGURE 6. The lock chamber of lift lock number 4 during excavation, looking northeast. The heavy, gray puddling clay used to line the lock chamber is clearly visible. *Photo: Ronald C. Carlisle.*

out-flaring wing walls at the downstream end of the chamber (see Figure 5). These walls were 5 feet thick at the base and 3 feet thick at the top.

Three feet up the wall from the base, a second timber foundation was placed on top of the stones. Perpendicular cross timbers were dovetailed into these timbers to form a cell that was then filled with rubble stone. The upper courses of the wall were then laid on top of the filled cell. The interior of the lock chamber was faced with a double row of pine planks nailed to oak stanchions keyed into the stone walls of the chamber.

Massive wooden gates blocked each end of the lock chamber. Paddle valves or wickets were built into the gates below the water line to let water in or out of the lock chamber. The gates in the lift lock were closed against a heavy wooden triangular construction called the miter sill. The gates rotated on an iron pintle in the floor of the chamber and folded back into a recess in the wall of the chamber when fully opened. A vertically implanted timber, called a quoin, engaged the rounded pivot edge of the gate, eliminating the need for metal hinges.

Only the lower 3 feet of the weigh lock, which stood parallel to and just south of the lift lock, were preserved as the floor of the weigh lock was some 9 feet above that of the lift lock. The weigh lock housed a huge steelyard scale used to determine the weight of the boat and hence its toll. After a boat had entered the weigh lock, the water in the lock was drawn off, and the boat came to rest in a large cradle attached to the scale. Unfortunately, no archaeological evidence of this scale or the weigh lock apparatus was preserved. This weigh lock was not a through-lock; rather, it opened only at one end, and the 19-foot-wide entrance was closed off by a single-span gate more common to dry docks than to canal locks. The gate pivoted on an iron gudgeon, and its far end was supported by a wheel that ran along a curved metal track.

Preservation, Conservation, and Reconstruction

Following the identification of the lift and weigh locks during the summer of 1987, a massive cooperative effort among Pittsburgh preservationists and government officials salvaged almost all of the stones, along with representative samples of the remarkably well-preserved wooden elements. The wooden structural components included one gate from the lift lock, the lock's miter sills, 37 structural timbers, and about 186 planks. Since the canal locks could not be preserved in place, a great deal of time and energy were devoted to negotiating plans for removal and storage of the stone, to selection and conservation of the wood, and to initiating a plan for future reconstruction and interpretation of the locks in celebration of Pittsburgh's canal days (Figure 7).

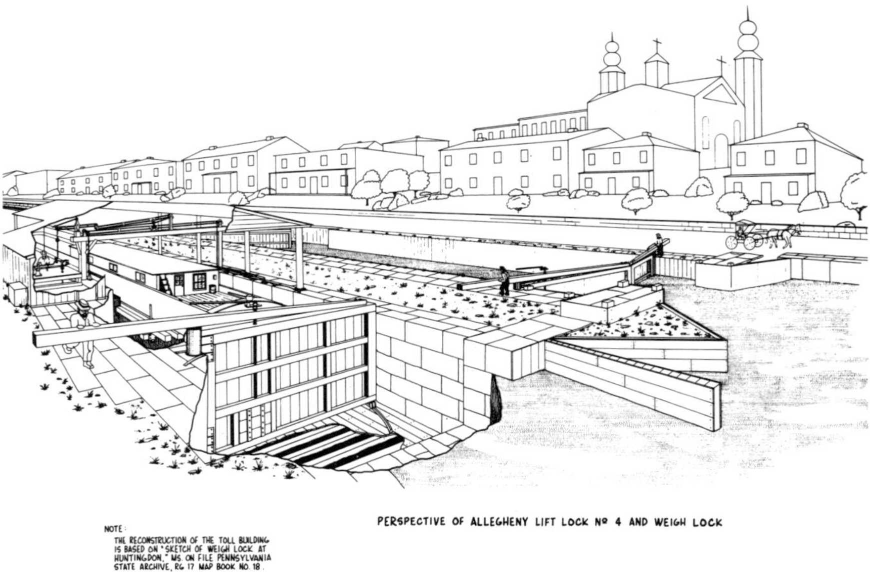

FIGURE 7. Artist's conception of the appearance of lift lock number 4 and the weigh lock showing a hypothetical reconstruction of the weighing mechanism. *Historic American Engineering Record, National Park Service. Delineated by Paul R. Pinchot, 1988.*

The Carnegie Museum of Natural History, the Committee on Pittsburgh Archaeology and History (CPAH), Pittsburgh's Department of City Planning and its Historic Review Commission, The Northside Civic Development Council, and The East Allegheny Community Council spearheaded the local effort to save the structures for future reconstruction. An Ad Hoc Canal Committee composed of members from these organizations worked with representatives from the Federal Highway Administration (FHWA), PennDOT, and the project's engineering consultants (Parsons Brinckerhoff Construction Services, Inc., and GAI Consultants, Inc.), to resolve what at times looked like insurmountable obstacles. Staff from the Bureau for Historic Preservation, Pennsylvania Historical and Museum Commission, also participated in the negotiations, and consultants from the Hugh Moore Historical Park and Museums in Easton, Pennsylvania, visited the highway-construction site and verified the importance of the salvage and reconstruction efforts.

The FHWA and PennDOT allocated $250,000 for the archaeological work. A skilled stone mason was given the contract to number, tag, and remove the stone to pallets. The City of Pittsburgh, through its Parks Department, provided secure, fenced space for the storage of the stones.

The Committee on Pittsburgh Archaeology and History and the Northside Civic Development Council requested and received a $25,000 emergency grant from The Pittsburgh Foundation to initiate conservation of the

FIGURE 8. Some of the 186 planks recovered from the canal lock excavations in Pittsburgh in 1987. These specimens as well as structural timbers from the site and one gate from the lift lock are being conserved for future use in a canal lock reconstruction to interpret the significance of the Pennsylvania Canal in Pittsburgh's history. Recovery of the wood has also stimulated interest in dendrochronology projects in western Pennsylvania. *Photo: Ronald C. Carlisle.*

wooden elements. Select architectural timbers and planks were then removed to the University of Pittsburgh's Applied Research Center (U-PARC) at Harmarville, Pennsylvania, for storage, conservation treatment, and study of the timbers under the professional guidance of Joan S. Gardner, archaeological conservator of the Division of Anthropology, The Carnegie Museum of Natural History (Figure 8; Carlisle et al. 1988).

The recovery of such a significant amount of wood from the canal locks also stimulated the development and implementation of a dendrochronology project by one of us (Carlisle), who also prevailed upon paleoethnobotanist Frances B. King and graduate student Renée M. Bonzani to participate. This was the first such project to be attempted for southwestern Pennsylvania and provided a topic for Bonzani's M.A. research paper in anthropology (Bonzani 1989; Bonzani et al. 1990, 1991). Eighteen samples of white pine (*Pinus* sp.) were analyzed using a tree-ring dating program to develop a "floating" chronology for the canal specimens. Eleven of the sampled planks were found to match at significant values to form a 152-year chronology. The floating chronology was then cross-dated with a white pine master chronology from Cook Forest, Pennsylvania. When cross-checked with other lines of evidence, the dendrochronological study showed that the trees used in constructing the canal locks had begun to grow by A.D. 1658. The outermost extant rings preserved in the pine sample were found to date to 1809.

More recently, a five-core sample taken from the oak structural timbers has produced a *tentative* date range from A.D. 1595 to 1811; however, replication in the floating chronology is poor, and additional study will need to be conducted before the tentative chronology can be confirmed (Robert Long 1990, pers. comm.). Utilizing information gained from the canal dendrochronology study, Carlisle and his co-workers extended this study to three historic structures in southwestern Pennsylvania. The purpose of the new work was to supplement the master regional chronology based on the canal specimens and, if possible, to provide more accurate construction dates for the structures.

While the wood conservation and dendrochronology studies were proceeding, other efforts were directed toward finding a suitable location for reconstruction of the locks, with a facility where the preserved wood and other objects could be used to interpret the city's canal history. CPAH, The Historical Society of Western Pennsylvania, and the two Northside community councils obtained a mini-grant from the Pennsylvania Humanities Council to sponsor a conference, "New Life for Old Locks: The Pittsburgh Canal Project" in the autumn of 1988. CPAH also established a Pittsburgh Canal Fund and encourages private contributions to the project.

Stimulated by the success stories of consultants who participated in this conference, the city of Pittsburgh and the Northside community councils launched searches for suitable locations where the locks could be reconstructed. After much discussion of several alternative sites, the city approved a parcel of land within a proposed river-park development on the Northside, very close to the point where the Main Line Canal crossed the Allegheny River via an aqueduct.

Once this site had been selected, CPAH and the East Allegheny Community Council submitted and received grants from the Pennsylvania Historical and Museum Commission and the National Trust for Historic Preservation to initiate a survey of the selected location and to generate a contour map of the parcel. The site study was completed in 1990. Eventually, the park may also house a 20-foot remnant of the Grant's Hill tunnel that carried the canal from the Allegheny to the Monongahela River. Pittsburgh's Department of City Planning will complete the remaining engineering studies to site the locks and the portion of the tunnel on the Canal Park site. The Northside community councils remain fully supportive of the project, and construction funds are currently being sought by them and by the city's Parks Department. CPAH continues to work on the reconstruction and interpretive exhibit portions of the project through the functioning of its own canal committee and through the contributions of its members who serve on the Ad Hoc Canal Committee that was such a vital force in saving the locks for the future.

The Ad Hoc Committee was recognized for its contributions to Pittsburgh's history when the canal project was nominated and received awards from the Historic Preservation Awards Program, U.S. Department of Transportation, Washington, D.C.; from the PennDOT Awards Program, and when it received the Pittsburgh History and Landmarks Foundation's Award of Merit.

References

Bonzani, Renée M.

1990 *Dendrochronology of the Pennsylvania Main Line Canal Lock Number Four, Pittsburgh*. Unpublished M.A. Thesis, Department of Anthropology, University of Pittsburgh, Pittsburgh, Pennsylvania.

Bonzani, Renée M., Ronald C. Carlisle, and Francis B. King

1990 "Dendrochronology of the Pennsylvania Main line Canal Lock Number Four, Pittsburgh." Paper presented at the sixty-first annual meeting of the Society for Pennsylvania Archaeology, Wilkes-Barre, Pennsylvania, April 27-29, 1990.

1991 "Dendrochronology of the Pennsylvania Main Line Canal Lock Number Four, Pittsburgh." *North American Archaeologist* 12(1): 61-73.

Burkett, Carl K., Jr., and Robert D. Ilisevich

1983 *A Survey of Historical Sources Relating to the Pennsylvania Main Line Canal in Pittsburgh*. Pittsburgh, Pennsylvania: Report prepared for the Urban Redevelopment Authority of Pittsburgh.

Carlisle, Ronald C., and Deborah Dalesandry

1988 "Historical Background." In *Phase I Archaeological Investigations at the Michael J. Kirwan Dam and Reservoir, Portage County, Ohio*, by B.J. Gundy, J.M. Adovasio, and D.R. Pedler. Harrisburg, Pennsylvania: Report prepared by the Cultural Resource Management Program, Department of Anthropology, University of Pittsburgh, for Gannett Fleming Water Resources Engineering, Inc.: 2.13-2.22.

Carlisle, Ronald C., Renée M. Bonzani, and Frances B. King

1991 "A Dendrochronology Study of Three Western Pennsylvania Historic Structures." Local History Project Support Grant Report. Submitted to the Pennsylvania Historical and Museum Commission, October 31, 1991.

Carlisle, Ronald C., Joan S. Gardner, and Nina M. Larsen

1988 "Conservation of Wood Recovered from Pennsylvania Canal Lift Lock Number 4, North Side, Pittsburgh, Pennsylvania." A paper presented at the "New Life for Old Locks" Symposium, annual meeting of the Committee on Pittsburgh Archaeology and History held in conjunction with the Northside Civic Development Council, the East Allegheny Community Council, and the Historical Society of Western Pennsylvania at the Buhl Science Center, Pittsburgh, Pennsylvania, October 15, 1988.

Humphrey, J. Steven

1990 Pennsylvania Local History Grant Application, Delaware and Hudson Canal, Pennsylvania.

Ilisevich, Robert D., and Carl K. Burkett, Jr.

1985 "The Canal Through Pittsburgh: Its Development and Physical Character." In *The Western Pennsylvania Historical Magazine* 68(4): 351-371.

Irion, Jack B.

1989 "The Excavation and Recordation of Lift Lock Number 4 and a Weigh Lock on the Pennsylvania Main Line Canal, Pittsburgh, Pennsylvania." A paper presented in the Industrial Archaeology Section of the Bureau for Historic Preservation Annual Meeting, Pittsburgh, Pennsylvania.

Lantz, Gary W.

1980 *A Cultural Resource Inventory of the East Street Interchange*. Report prepared by the Section of Man, Carnegie Museum of Natural History, Pittsburgh, Pennsylvania, for the Pennsylvania Department of Transportation.

Leach, Sarah Amy, editor

1989 *Two Historic Pennsylvania Canal Towns: Alexandria and Saltsburg*. Washington, D.C.: U.S. Department of the Interior, National Park Service, Historic American Buildings Survey/Historic American Engineering Record.

McCullough, Robert, and Walter Leuba

1962 "The Pennsylvania Main Line Canal." *Morrisons Cove Herald*, Martinsburg, Pennsylvania.

Reid, R.D.

1983 *Two Hundred Years of Water Transportation in Eastern Ohio and Western Pennsylvania*. Talmadge, Ohio.: R.D. Reid.

Rhoads, William R.

1960 "The Pennsylvania Canal." *The Western Pennsylvania Historical Magazine* 43(3), 203-237.

Scheiber, Harry N.

1969 *Ohio Canal Era: A Case Study of Government and the Economy, 1820-1861*. Athens, Ohio: Ohio University Press.

Shank, William H.

1984a "Those Amazing Pennsylvania Canals," Part 1. *Pennsylvania Magazine* 3(1): 36-41.

1984b "Those Amazing Pennsylvania Canals," Part 2. *Pennsylvania Magazine* 3(2): 51-56.

SANTEE CANAL—SUCCESS OR FAILURE?
A Preliminary Evaluation of the Economic Impacts of the South's First Major Transportation Venture

Mark M. Newell

Abstract

This article makes a preliminary evaluation of the direct and indirect evidence for the impact of the building and operation of the Santee Canal (1786 to circa 1860) on communities in antebellum South Carolina, particularly Charleston. The purpose of the venture was to build the first true summit canal in North America, and it proved popular with South Carolina's planters and Charleston's merchants alike. Like many other early American canals, the venture was a dismal failure for its investors, but it was a valued and successful development for the state's merchant and planter communities during the brief period when no other viable transportation alternatives were available.

Introduction: Great Expectations

South Carolina's post-Revolutionary War years were marked by efforts to recover from the economic depression that followed the years of conflict with Great Britain's forces. A direct impact on the fortunes of the state was the damage and disruption caused to the plantation system by the war. Plantations lost labor forces, animal stock, and equipment (Orvin 1973: 14). An indirect but equally serious impact was the loss of British markets for the products the plantations produced: rice, naval stores, and indigo (Orvin 1973: 147).

Leading figures in South Carolina society began to work for the revitalization of the plantation system, and in 1785 the state's "delivery system" between major plantations and Charleston, the major port of the era, became a focus of this activity (*South Carolina Gazette* November 12, 1785). Prior to the war, South Carolina's burgeoning plantation system relied on natural waterways and a variety of watercraft (Newell 1986: 1) to deliver cargoes to Charleston and return imported supplies and finished goods. The vessels designed to negotiate river systems were ill suited to survive the ocean voyages from coastal estuaries to Charleston. The rivers themselves meandered over coastal lowlands creating time-consuming and tortuously indirect routes that added to the time and expense involved in delivering a cargo to Charleston.

FIGURE 1. Staff members of the South Carolina Institute of Anthropology and Archaeology inspect a circa 1845 cypress tidelock gate found in Biggin Creek, South Carolina, below the terminus of the Santee Canal. *Photo: Mark M. Newell.*

The Santee River and its upland tributaries were prime examples of this problem. The river provided a coastal route to Charleston for much of central upland South Carolina and waterways that even extended into North Carolina. In 1785, a group of entrepreneurs, merchants, and plantation owners met in Charleston to seek a legislative charter to incorporate The Santee Canal Company (Porcher 1950: Appendix).

Inland navigation had been introduced into England by the Earl of Bridgewater in the 1760s (Rolt 1973: 29), and by this time canal technology had been developed to a high level of sophistication. Efforts to import the technology into North America had been occurring for some time — Benjamin Franklin was himself a prime proponent (Ringwalt 1966 [1888]: 41-42).

The canal project proposed by the Santee Canal Company was the single most ambitious effort of the era. At the end of the 18th century, navigation projects had attempted no more than the circumvention of falls and shoal areas (Garrett 1987: 716). The Santee Canal, by contrast, was to be a 22-mile canal joining the Santee River to the headwaters of the Cooper River by a series of rises and falls connected by a 5-mile summit.

The composition of the stockholders of the canal company—bankers, planters, merchants (Jaher 1982: 350), and even early industrialists (Cowan 1987: 6)—supports the generally accepted view at the time that "great economic advantage" (Crowson 1971: 7) would be gained from the operation of the canal. Even so, it took until 1793 for the company to establish its base of operations at a plantation called Big Camp (Bennett and Richardson 1988: 168), acquire land for the route (Porcher 1950: Figure 2), and commence the digging of the canal (Crowson 1971: 8).

Economic Impacts of Construction

Five routes for the Santee Canal utilizing natural waterways had been surveyed and recommended as early as 1772 by Henry Mouzon (Porcher 1970 [1903]: 3). The engineer selected in 1787 for the project, Col. Christian Senf, chose a sixth route running from White Oak Plantation on the Santee River through the Biggin Swamp Basin to Biggin Creek at the head of the Cooper River (Porcher 1950: Figure 1). White Oak was owned by one of the largest canal company stockholders and planters, Ralph Izard. Izard planned to benefit from the location by developing Izardton at White Oak — giving rise to the claim by Senf's critic, F.A. Porcher (Porcher 1970 [1903]: 4), that the engineer had been influenced by Izard's greed in his choice of route. Porcher, a failed cotton planter, like other proponents of the Mouzon routes, did not recognize the fact that sediment-filled water from natural watercourses were to be avoided at all costs (Rolt 1973: 34).

The immediate gains from sale of real estate were followed by the profitable hiring out of slaves by local planters to the canal company for the digging operations. Initial rates were £15 to £16 per year per slave (Webber 1954: 120), the number of slaves engaged in the project rapidly rising from 10 in May of 1793 to more than 1,000 at the end of the year (Crowson 1971: 8). The base of operations for the massive project was "Big Camp", the name given to a plantation purchased by the canal company specifically for this purpose (Bennett and Richardson 1988: 87-88). To the work force needed for digging the canal were added the craftsmen needed to fashion the canal's architectural components. A large number of blacksmiths were early employees of the company (Crowson 1971: 9), and the company advertised in the *South Carolina Gazette* for "negro bricklayers." A total of 25 carpenters were also hired or employed by the company to work on the ponderous conversion of lumber cleared from the swamps into lock gates, foundation beams, boards, and other construction elements of the canal (Crowson 1971: 9). Twenty-five skilled stonemasons were also employed to carve the many hollow quoins and the decorative stonework used throughout the canal (Newell 1989: 94-95). To what extent these craftsmen were slaves, freedmen, or indentured whites is not known. Clearly, at least some of the brick masons

FIGURE 2. Modern artist's enhancement of an 1835 pencil sketch of Black Oak Lock on the Santee Canal. *Contemporary drawing: Christine Madigan.*

were black. It is also known that some of the blacksmiths were black (Deas 1941: 27). Stonemasons, in contrast, were highly skilled whites. They were few in number and usually travelled widely to ply their craft (Newell 1989: 94). It is therefore likely that craftsmen were a third group that economically benefitted from the actual canal construction process after land and slave owners. No mention is made of another cost/profit aspect of the project, food supply, other than the fact that 34 cooks were required to prepare the food.

Canal Operations: Economic Impact on Stockholders

The expectations of stockholders were dampened long before the Santee Canal opened for business. As the work progressed, costs soared. Prices for slave labor increased to £24 per annum for men and £20 for women. Overseers' wages increased from £30 to £60 over the same period (Webber 1954: 120-124). Added to these immediate costs were the costs of delays in the work. The first three years of the work suffered delays due to unseasonably heavy rains (Simmons and Newell 1988: 19). Finally, there remain questions about the manner in which Senf managed the project. He is said to have been overly meticulous and too personally involved in the details of the construction (Porcher 1950: 4). The original cost of the project was estimated at £55,620 (Webber 1954: 119). According to Senf, the final cost by 1800 had amounted to twice that figure (Webber 1954: 119). The original completion date for the canal had been projected as 1796 (Bennett and Richardson 1988: 42). Later, dollar estimates of an "ultimate" cost including the principal and interest paid on funding the project ranged from

Table 1

Initial Earnings of the Santee Canal From Tolls*

Year	Earnings
1802	$3,083
1803	$5,407
1804	$6,886
1805	$10,563
1806	$12,776
1807	$11,288
1814	$21,249

*Annual estimated maintenance cost is $7,000.

Sources: Crowson 1971: 14; Bennett and Richardson 1988: 65.

$1,100,000 to $1,615,000 (U.S. Army Corps of Engineers 1881: 1152). As these costs mounted, the original stockholders were directly liable for the increased cost of each share, and some of them were financially crippled by these demands (Porcher 1950: 10). It was the opinion of many that the cost was such that no reasonable toll rate for traffic on the canal would ever return the investment (Phillips 1908: 41).

The company's founders proposed an initial stock issue of 1,000 shares valued at £100 each. Maximum tolls were set at a rate to earn no more than four times the annual cost of maintenance of the canal (Bennett and Richardson 1988: 4). Initial earnings from tolls were promising (Table 1). Tolls were first set at a single rate, $21.00, irrespective of vessel size and capacity. The rapid increase in receipts in 1805 was most likely due to the introduction of a sliding scale for vessels based primarily on beam and capacity (Table 2). How long the sliding scale of tolls was in use is not known. By 1821 a single toll figure of $15 per passage is mentioned (Phillips 1908: 43). This is not, however, the first indication of problems in the canal's operations and funding. Starting in 1817, there was a three-year drought of such severity that even the natural waterways surveyed by Mouzon were dry (Bennett and Richardson 1988: 47). In 1818 the directors of the company petitioned the state legislature for an interest-free $30,000 loan for two steam pumps to supply the canal's summit with water (General Assembly Petitions 1818: 45). The petition gives an assessment of the canal's operations to that date: "Your petitioners are free to admit that the advantages to the public have not as yet been so great as were at first intended...." The petitioners continued, indicating that efforts to make the canal "...yield some return to the original undertakers..." had totally failed. They also stated that "...the expenses of this canal have transcended the estimates in a degree unprecedented, and to many of the Stockholders, ruinous."

Table 2

Capacities of Vessels on the Santee Canal, Number of Crew and Tolls Paid

Vessel type	Length (ft..)	Beam (ft.)	Draft (Inches) Empty	Draft (Inches) Loaded	Capacity[a]	Crew	Toll
Canal boat	56	9.5	24	36	80-120	5-6	$20
Canal boat (2)	56	8.5	12	24	80	4-5	$18
Mountain boat	56	7.5	8-10	18-24	70	3-4	$14
Mountain boat (2)	56	6.5	8-10	14-17	50	3	$10
Flat boat	54	9.5	8	24	70	5	$30
Flat boat (2)					120-130		$60[b]

[a]Capacity is expressed in number of cotton bales.

[b]Given as $35 by Crowson (1971:9). Vessels of capacity over 70 bales were subject to a toll surcharge.

The weather was not the only factor that affected profits. Embargoes and "nonintercourse" policies of the federal government during the War of 1812 cut down traffic on the canal. The company was required by its charter to keep the canal open during this period, but receipts did not cover costs and repairs. True, there had been a brief economic revival after the War of 1812, resulting in the highest recorded earnings from tolls (see Table 1). In 1814, the company's net profits were $12,389, and in 1815 they were $13,086 (Bennett and Richardson 1988: 65), although toll earnings for 1815 are not given in the same reference. The petition resulted in further studies on the viability of the canal by then-State Engineer Abram Blanding and engineer/canal stockholder Henry Ravenel. Their report recommended deepening the summit and widening the locks to allow riverboats and steamships with towboats through the canal. The improvements were not financed, however, and the steam pumps failed to draw sufficient water from the Santee River to supply the summit.

The drought years were disastrous for the canal, and they were followed by (and doubtless contributed to) the depression of 1819, despite the fact that 1818 was an $11 million record year for export crops from South Carolina (Pease and Pease 1985: 41). Other factors also began to affect canal earnings. The reduced toll of 1821 may also have been a response to growing competition from steamboat traffic. By 1822 steamboats had been developed to the point where they could carry 900 to 1,000 bales of cotton from upland markets to Charleston in half the time it took a smaller vessel to navigate the Santee Canal (Bennett and Richardson 1988: 69; Crowson 1971: 16). Conditions improved somewhat as navigation projects farther inland were developed on upland rivers north of the canal (Bennett and Richardson 1988: 57), but business during the next decade or so was still depressed by low international prices for cotton and rice. The depressed prices also had the effect of reducing imports that might otherwise have been shipped through the canal (Pease and Pease 1985: 41). New national tariffs designed to protect northern manufacturing interests also had an effect after their introduction in 1828 and 1832, and South Carolina's intense preoccupation with attempts to nullify those tariffs caused the economy to languish (Pease and Pease 1988: 42). During this period real estate prices in Charleston plummeted, homes and businesses stood empty, and banks refused to advance loans.

The political crisis had been resolved by 1833, and South Carolina embarked upon a period of recovery. For the next six years cotton and rice exports from Charleston topped $10 million annually (Pease and Pease 1985: 49). Much of this improvement was due to increased crop production (Pease and Pease 1985: 280, Note 15). More crops also increased demand for more transportation routes to Charleston. The railroad soon became a competitor with the canal (Jaher 1982: 350-351), and the number of arrivals of upland cotton in Charleston by river increased so much that a new wave of speculation in wharfage ownership was created (Pease and Pease 1985: 49-51).

Little of this traffic appears to have funnelled through the Santee Canal. There is some confusion in the published record as to dividends paid to stockholders during this period (Table 3). Orvin (1973: 153) states that a $12.25 dividend was paid in 1809 and $10.00 in 1810. This contrasts with a statement made by stockholders in 1853 to the effect that no dividends were paid until 1816 (Bennett and Richardson 1988: 70). Returns apparently peaked in 1828 when semiannual dividends were $20.00, this during a period of economic recession. By 1837, when recovery was well underway, semiannual dividends had actually dropped to $8.00 per share (Orvin 1973: 153). Crowson (1971: 17), without referencing his source, states in contrast that a dividend averaging $27.00 was paid from 1825 to 1841

Table 3

Dividend History of the Santee Canal

Year	Dividend Type	Total Dividend
1800-1808	—	$-0-
1809	Annual	$12.25
1814	Annual	$10.00
1820	Annual	$45.00
1828	Semi-annual	$40.00
1837	Semi-annual	$16.00
1841	(stock subdivision into 12 units)	—
1841-1849	Semi-annual	$1.28*
1850-1853		$-0-

*$15.36 in pre-1841 stock.

Sources: Orvin 1973; Bennett and Richardson 1988; Crowson 1971; Porcher 1950.

(Crowson 1971: 17). Orvin relates that the semiannual dividends continued to decline until they reached $ .50 in 1845. Using General Assembly petitions as their source, Bennett and Richardson (1988: 71) report dividends averaging $1.28 from 1841 to 1849.

Within the next decade, when the value of Charleston's exports and imports rose 71 percent (Jaher 1982: 353), the stockholders of the Santee Canal were petitioning the legislature for permission to close it (Bennett and Richardson 1988: 70). The petition was submitted and granted in 1853. This period was not one of great railroad growth (Jaher 1982: 353); it was the steamboat that was apparently filling the gap.

It already has been noted that there was a temporary benefit to the region's war-devastated plantation owners in the opportunity to hire out slave work forces for the construction of the Santee Canal. As the annual price of labor indicates (Webber 1954: 120), however, slaves were more profitable back in the plantation fields as the 18th century came to a close. This was due in part to the introduction of cotton in 1795 (Porcher 1950: 4). By this time, agricultural produce was being harvested as far inland as Tennessee (Simmons and Newell 1988: 9). In South Carolina, certainly, most of these plantations were within a few miles of a navigable waterway. Four major upstate rivers, the Saluda, Broad, Catawba and Fishing Creek, provided access from these areas via the Congaree and Wateree rivers. Thus, more than two thirds of the upland plantations, in addition to Tennessee and North Carolina markets, used the Santee River watershed as a primary means of transportation to Charleston before the construction of the Santee Canal. As Porcher (1950: 1) notes: "It is impossible to over-estimate the importance of the Santee River as a channel of commerce." The principal difficulty with all river transportation not directly connected with Charleston harbor was the time involved in traversing the Coastal Plain and the inevitable sea journey along the Carolina coast.

Not everything traveled by river, however. An early network of postroads traversed the state and were traveled by large-load wagons and mule teams (Newell 1989: 14). In England, where comparisons of early modes of travel had already been made (Rolt 1973: 1), the figures given in Table 4 prevailed. It is reasonable to assume that they also apply to similar transportation modes in South Carolina at the same time.

A Senf contemporary, Dr. Charles Drayton, made frequent visits to the canal construction project. On May 6, 1794 he recorded (Drayton 1796) for the Santee canal boats:

> Boats carry 18 tons or 720 bushels of corn, or 200 barrels of flour, or 35 h[ogsheads] of tobacco, navigated by one boy and 4 men. This is the same amount of work as 18 wagons, 36 men and 90 horses.

Table 4

Load Comparisons of Various Transportation Modes

Average pack-horse load	⅛ ton
Stage wagon, soft road	⅝ ton
Stage wagon, surfaced road	2 tons
Barge on river	30 tons
Barge on canal	50 tons
Wagon on rails	8 tons

Source: Rolt 1973

Drayton seems to be referring to the larger flats that traveled the canal since his 18-ton figure concurs with the 120-cotton-bale figure quoted in Table 2, given that the average cotton bale weighed 300 lbs (Crowson 1971: 17).

When the Santee Canal opened, the major problem of transporting these types and weights of commodities to Charleston was largely solved. The river transportation trade of the antebellum period was conducted by plantation owners, by independent ship captains (Nylund 1989: 104), mountain boatmen (Terrell 1988: 53), and merchant factors. Plantation owners who built their own craft also certainly used the canal (Porcher 1950: Appendix). When the canal opened, the canal company also participated in this trade, building and operating its own canal craft (Webber 1954: 112).

Unlike the financial fortunes of the company's stockholders, it is far more difficult to find statistics that reflect the economic impacts of the canal on the planting, transportation, and merchant communities. That they did benefit may be confidently inferred from sources of "peripheral information."

The anticipation of great economic returns is certainly evident in the statements made in support of the project prior to its construction. Charlestonians, who suffered the same economic malaise as did those on the post-Revolutionary War plantations, saw the canal project as the cure: "Many Charlestonians felt that if the Santee could be connected directly with their city, a great economic advantage would be enjoyed." (Crowson 1971: 7). This optimism was reflected in the make-up of the board of directors of the canal company and its shareholders in 1786 (Porcher 1950: Appendix), many of whom were important urban politicians, planters, bankers, merchants and debt speculators (Jaher 1982: 350). During the period from 1828 to 1832, most of the company's directors were either wealthy merchants or factors; the balance, about 14 percent, were planters (Pease and Pease 1985: 279, Note 5).

In June 1801, approximately one year after the canal opened, *The Times* of Charleston reported on the arrival of a plantation-built vessel in

Charleston from Pinckney Court House, 90 miles above Columbia. Noting the distance traveled, the paper commented that the arrival: "...leads to the most pleasing anticipation of the immense advantages which are likely soon to result from the easy intercourse that may now be carried on, between Charleston and the interior country, by means of the Santee Canal" (quoted in Porcher 1950: Appendix).

This report also said of the canal company stockholders that while the canal would contribute to the welfare of the state it would also: "...yield them an income that will amply repay all that they have expended in an undertaking which is unequalled in the new world."

Porcher's history of the canal, despite its inaccuracies and bias, properly credits the canal with playing a significant role in Charleston's early-19th-century development. He commented incorrectly, however, that the Charleston Library Society's newspaper files reveal nothing of the activities of the canal company prior to 1820: "Until the year 1820 one would never learn from our papers that the commerce of this city *was largely fed by a canal, which made access to the upper country comparatively easy*." (Emphasis added; Porcher 1950: 1). Porcher (1950: 1) also makes another significant observation in his comments on the effects of the canal's closing during the droughts of 1817, 1818, and 1819: "Then were all the roads to Charleston whitened by the tents of the wagons as they slowly pursued their toiling way to and from the metropolis."

Direct correlation between statistical information for this period and the value of the canal is difficult to find. Clearly the canal played a role in the economic growth of Charleston, and that growth was significant during the first decades of the 19th century as indicated by Tables 5 and 6.

Population growth may also be taken as an indicator of economic growth. Especially significant is the growth in slave population as this segment of the community would reflect growth of the labor force in response to increased commercial and agricultural activity (Table 7).

Table 5

Value of Exports from the Port of Charleston for the Years 1799-1825

Year	Value
1799	$10,544,842
1800	$14,304,045
1824	$7,143,831
1825	$16,000,000

Sources: Mills 1826; Jaher 1982.

Table 6

Government Revenues Collected from the District of Charleston, 1800-1825

Year	Amount
1800	$812,698.99
1801	$1,119,688.29
1802	$260,909.37
1803	$639,656.51
1804	$707,053.33
1805	$837,849.49
1806	$665,085.14
1807	$728,382.56
1808	$309,685.90
1809	$378,778.63
1810	$566,569.66
1811	$334,869.32
1812	$429,179.93
1813	$251,844.96
1814	$126,046.68
1815	$1,408,863.41
1816	$1,356,927.87
1817[a]	$1,616,460.75
1818[a]	$1,245,980.75
1819[a]	$734,758.58
1820	$549,866.04
1821	$511,852.90
1822	$780,470.21
1823	$735,000.00
1824	$741,038.00
1825	$701,003.00

[a] Drought years when wagons predominated as transportation mode. Source Mills 1826.

As can be seen from the Charleston comparison for 1825, almost 50 percent of the free Black population and 10 percent of the slave population for the state was located in Charleston. These figures certainly do not allow any definitive statements to be made about the impact of the canal. It can only be said that the city's economy expanded in the early 19th century, a period when at least one source claims that the Santee Canal was "largely" responsible for its commerce (Porcher 1950 1).

Table 7

Population Growth of South Carolina, 1790-1820

Year	White	Slave	Free Black
1790	140,178	107,094	1,801
1800	196,255	146,151	3,185
1810	214,196	196,365	4,554
1820	237,440	258,475	6,826
Comparison figures for City of Charleston:			
1820 (Total figure only: 37,471)			
1825	17,000	20,000	3,000

Source: Mills 1826

Conclusion

Clearly, the Santee Canal Company was a dismal failure as an investment. Its value to the plantation and merchant communities seems clear, but it is not yet provable. It also contributed to the urban growth of Charleston, but to what extent and in what exact ways it is not possible to say on the basis of currently available data. The canal operated for over 50 years, and its closure was resisted (Bennett and Richardson 1988: 72-75). These are further indications of its value to these communities. Documents bearing on the canal operations are known to still be the hands of descendants of stockholders. If the information contained in those documents becomes available for study, more satisfying proof of the Santee Canal's value to the development of South Carolina and Charleston may be able to be published.

References

Bennett, Robert

1988 *The Santee Canal 1785-1939*. Unpublished M.A. Thesis, Department of History, University of South Carolina, Columbia, South Carolina.

Bennett, R., and K. Richardson

1988 *The History of the Santee Canal and its Environs, 1785-1900*. Report prepared for the South Carolina Department of Parks, Recreation, and Tourism. Report on file at Caroliniana Library, University of South Carolina, Columbia, South Carolina.

Cowan, T.

1987 "William Hill and the Aera Ironworks." *Journal of Early Southern Decorative Arts* 13(2): 1-31.

Crowson, E.T.

1971 "The Santee Cooper: A Pioneering Effort in South Carolina Canaling." *The South Carolina Magazine* 35(8): 6-9.

Deas, A.

1941 *The Early Ironwork of Charleston*. Columbia, S.C.: Bostick and Thornley.

Drayton, Charles

1796 *Diary of Charles Drayton*. Unpublished manuscript diary on file at the Charleston Historical Society, Charleston, South Carolina.

Garrett, Wilbur E.

1987 "Waterway that Led to the Constitution: George Washington's Patowmack Canal." *National Geographic Magazine* 171(6): 716-753.

Jaher, F.C.

1982 *The Urban Establishment*. Chicago: University of Illinois Press.

Kohn, D., compiler and editor

1938 *Internal Improvements in South Carolina, 1817-1828*. Washington, D.C.: Reports of the Superintendent of Works.

Mills, Robert

1826 *Statistics of South Carolina*. Charleston S.C.: Hurlbut and Lloyd.

Newell, M.M.

1986 "Current Small Craft Research in South Carolina." Ms. on file, South Carolina Institute of Archaeology and Anthropology, University of South Carolina, Columbia, South Carolina.

1989 *The Santee Canal Sanctuary, Part II: Preliminary Archaeological Investigations*. Cultural Resource Management Publication 6. South Carolina Institute of Archaeology and Anthropology, University of South Carolina, Columbia, South Carolina.

Nylund, R.C.

1989 *The Historical Background of the Brown's Ferry Vessel*. Unpublished M.A. Thesis, Department of History, University of South Carolina, Columbia, South Carolina.

Orvin, M.C.

1973 *Historic Berkeley County, South Carolina, 1671-1900*. Charleston, S.C.: Comprint.

Pease W.H., and J.H. Pease

1985 *The Web of Progress*. New York: Oxford University Press.

Phillips, U.B.

1908 *A History of Transportation in the Eastern Cotton Belt to 1860*. New York: Columbia University Press.

Porcher, F.A.

1950 *The History of the Santee Canal*. South Carolina Public Service Authority; reprint of 1903 edition. Charleston, S.C.: South Carolina Historical Society.

1970 *The History of the Santee Canal*. Berkeley County Tricentennial Committee, 1903. Reprint, Charleston, S.C.: South Carolina Historical Society.

Ringwalt, J.L.

1966 *A History of Transportation Systems in the United States*. Reprint of 1888 publication. London: Johnson Reprint Company Ltd.

Rolt, L.T.C.

1973 *Navigable Waterways*. London: Arrow Books.

Simmons, J.J., and M.M. Newell

1988 *The Santee Canal Sanctuary Part I: Preliminary Archaeological Survey*. Cultural Resource Management Publication 4. South Carolina Institute of Archaeology and Anthropology, University of South Carolina, Columbia, South Carolina.

Terrell, B.G.

1988 "The James River Bateau: Tobacco Transport in Upland Virginia 1745-1840." Unpublished M. A. thesis, Department of History, East Carolina University, Greenville, North Carolina.

U.S. Army Corps of Engineers

1881 *Survey for Re-Opening the Santee Canal. Report of the Chief of Engineers*. Appendix J 17: 1148-1157.

Webber, M.L., editor

1954 "Colonel Senf's Account of the Santee Canal." *The South Carolina Historical Magazine* 55: 8-21, 112-131.

THE RESURRECTION OF THE ALEXANDRIA CANAL TIDAL LOCK AND THE ROLE OF THE CANAL IN THE CITY'S PAST AND PRESENT

Steven J. Shephard

Abstract

The Alexandria Canal, with its four locks and basins, was completed amid great rejoicing by the citizens of Alexandria, Virginia, in 1845. This waterway connected the Chesapeake and Ohio Canal at Georgetown with the port of Alexandria on the Potomac River. Coal, wheat, whiskey, corn meal, and flour were shipped to Alexandria from the west. Fish, salt, plaster, and lumber were shipped from Alexandria to the west. The canal expanded the commercial horizons of Alexandria businessmen but also served as a means of trade for slaves and free Blacks, a convenient location for preaching the gospel, and an ice rink for skating. The canal ceased operations during the Civil War, underwent a brief revival during the 1870s, and was abandoned in 1886. Alexandria Archaeology rediscovered Lift Lock No. 1 (the Tidal Lock) in 1979. Following archaeological excavation, the lock was restored as part of an office complex development in 1986.

Introduction

The Alexandria Canal has been many things to many people. It was to be the salvation of Alexandria, Virginia's, dwindling trade in the minds of the citizens in the early 19th century, and it lives on as a hidden, but not forgotten, remnant of the past to Alexandrians of the 20th century. Numerous papers and articles have been written on the Alexandria Canal, most from the perspective of the history of its conception, financing, construction, and eventual abandonment. These aspects of the canal will be briefly addressed, but the thrust here is on economic and social effects as well as providing a glimpse of the breadth of the canal's impact both in the past and the present.

The history of the Alexandria Canal is a roller coaster ride of ups and downs—in soaring hopes of commercial success; in headaches caused by floods, droughts, and repairs; and in healthy profits from coal shipments. Concern for improvement of commercial navigation on the Potomac River began in the mid-18th century. By 1785, the Potomac Company had been organized with George Washington as its president. The goal of this company was to improve navigation by the construction of a series of short canals to bypass the falls above Georgetown, Maryland. Locks were to be constructed at Little Falls and Great Falls; boats could be poled the rest of the way downriver (Mitchell 1981: 4). A set of locks was built at Little Falls in 1795, and the Great Falls Skirting Canal was completed in 1802. The Potomac Company, however, went out of business in 1828, and the charter and rights of this organization were assumed by the Chesapeake and Ohio Canal Company (referred to hereafter as the C&O).

By this time it was clear that the difficulty of navigation on the Upper Potomac could be resolved only through the construction of a cut canal parallel to the river. An estimate for the construction of a canal from Cumberland, Maryland, to Georgetown was $4,500,000. The City of Alexandria, with hopes of realizing a major benefit from the project, subscribed $250,000 and reserved the right to construct a lateral canal to connect Alexandria with the C&O (Hahn 1984: 34). Each section of canal was used as soon as it was completed, but due to various difficulties it did not reach Cumberland until 1850—eight years after the Baltimore and Ohio Railroad had been completed to that town (Hahn 1984: 36).

Historic Overview of the Alexandria Canal

The Alexandria Canal Company was created in 1830. It was organized by Alexandria businessmen who had experienced a steadily declining economy for over a decade. During Alexandria's commercial heyday in the late 18th century, the sale and shipment of goods, especially tobacco and flour, formed the basis of the city's economy. Manufacturing was at a small scale. In 1817 the prices for agricultural commodities began to decline and continued to be depressed throughout the 1820s (Sharrer 1977: 23). It was expected that with the completion of the C&O Canal, Georgetown would become the terminus for western trade and a major port for the shipment of coal. Alexandria businessmen saw the Alexandria Canal as the means for tying into this trade route and supplanting Georgetown as the trans-shipment terminus. Building the canal became the answer to Alexandria's economic woes, and in the minds of the city's leaders, the only hope for the port's survival. Groundbreaking for the canal on the Fourth of July, 1831, witnessed a full day of organized celebrations, including salutes of musketry and a parade composed of a marching band; board members of the Alexandria Canal Company and the C&O Canal Company; the mayors and city councils of Alexandria, Georgetown, and Washington; members of at least five military and civil organizations; school teachers and their pupils; and lastly, strangers (*Alexandria Gazette* [AG] 6 July 1831).

The most challenging aspect of building the canal was the construction of the aqueduct over the Potomac River at Georgetown. Actual costs overran the estimate, and the Alexandria Common Council petitioned Con-

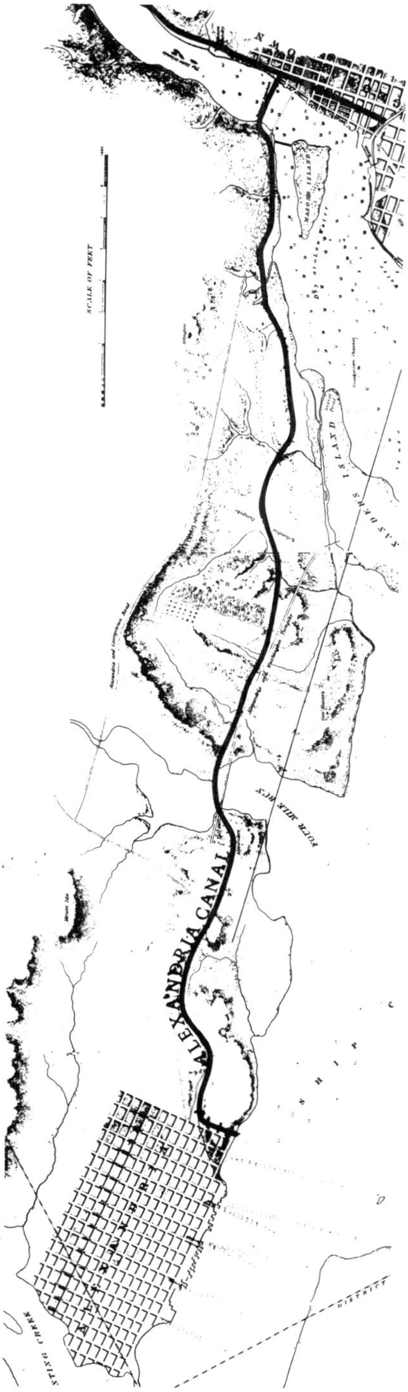

FIGURE 1. "Detail of Chart of the Head of Navigation of the Potomac River Shewing [*sic*] the Route of the Alexandria Canal." *U.S. Congress 1841.*

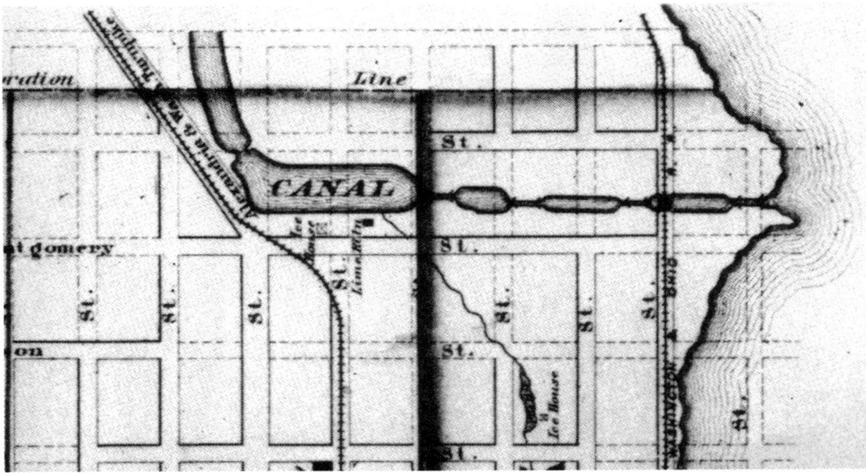

FIGURE 2. Location of the Alexandria Canal Turning Basin and four lift locks. *Hopkins 1877.*

gress for funding. The dire consequences for the future of Alexandria if the canal was not completed were emphasized in this memorial. The completion of the canal, it was said, was "of vital importance to the existence of their town," but due to various economic difficulties "the confidence of our people was almost destroyed; and that, in short, without the aid of Congress, utter and irretrievable ruin must shortly overtake them" (Corporation of Alexandria 1836: 1). Funds were obtained, and the work on the canal continued.

Despite numerous delays and difficulties in building the canal, enthusiasm ran high. An article in the *Alexandria Gazette* in 1839 declared:

> As the main work [the C&O Canal] is thus successfully carried on, we are rejoiced to repeat that the lateral canal which is to bring it to its natural and proper termination, is also pushing forward with commercial zeal. The prospect opens and brightens before us daily. Alexandria must put forth all her energies, so as to finish her Canal at the time the main Canal is finished to Cumberland. For the next two years go ahead! must be the word—and after that we are sure to go ahead! (*AG* 26 April 1839: 2).

The canal was completed from Georgetown to Alexandria in 1843 (Figure 1). A magnificent aqueduct, an engineering feat for its time, diverged south from the C&O and spanned the Potomac River from Georgetown to the Virginia shore. The canal continued on a level for 7 miles and ended in a 50 foot x 500 foot turning basin at the north end of Alexandria.

The arrival of the first canal boat, the *Pioneer*, on December 2, 1843, was cause for great celebration. The president and directors of the Alexandria

Canal Company embarked on the boat at Georgetown and were pulled by mule to Alexandria, where they were greeted by cheers from a large crowd gathered at the basin. A salute was fired, and the vessels in port were decorated with flags (*AG* 4 December 1843: 2). At this time, construction of the four locks needed to lower boats from the level of the turning basin to the Potomac was underway. The terminus of the canal was Lock Number 1, the Tidal Lock (Figure 2). A large wharf was located at the mouth of the lock so that canal boats could both discharge and load cargoes from transport vessels.

The canal was an immediate success. In the early years of operation trade flourished in agricultural products, especially wheat and corn, while the cargoes going back up the canal from Alexandria were manufactured products and fish (Hahn 1984: 53). By 1850, the C&O Canal was completed to Cumberland, Maryland, and from then on coal was the major product transported to Alexandria. During the Civil War, the aqueduct was drained and planked-over to form a bridge for federal troops and supplies crossing the Potomac River. The canal was thus abandoned for four years. Canal trade resumed after 1865 and continued with some success, thanks to the demand for coal, until about 1875 when canal shipping began a steady decline. In 1886, a severe break in the aqueduct brought an end to the Alexandria Canal. There was great public demand for a toll-free bridge across the Potomac River, and in 1886 the federal government purchased the aqueduct for conversion to this purpose (Hahn 1984: 62-63). The sides of the canal eroded over the years, while the locks and basins were used as receptacles for trash and industrial waste. In 1894, a complaint was published in the *Alexandria Gazette* about the skinned carcasses of animals buried near the Tidal Lock having been unearthed by dogs and so becoming "offensive and a menace to health" (*AG* 11 May 1894: 3). During the first quarter of the 20th century the Old Dominion Glass Factory used the Tidal Lock as a dumping area for the waste glass from its operation.

Eventually, the canal was entirely covered over and forgotten by many of Alexandria's residents. In 1974, Thomas Hahn, industrial archaeologist and then president of the American Canal Society, wrote a letter to the *Washington Post* (6 July 1974: A15), reminding the citizens of Alexandria of the significance of the canal and the possibility of the preservation of the Tidal Lock. His article prompted members of the Alexandria Archaeological Commission and a member of the City Council to learn more about the canal's history. In 1978, Alexandria Archaeology conducted test excavations and located the walls of the Tidal Lock, and in 1981, it was placed on the National Register of Historic Places. The following year, a grant from the National Trust for Historic Preservation provided funding for Hahn to excavate portions of the Tidal Lock and the adjoining basin. This work determined the location and condition of these features. Hahn's report on this investigation (Hahn 1982: 72-75) concluded by discussing the costs and consequences of six alternatives for preservation management of the site, ranging from no preservation action to full restoration and interpretation.

In 1984, Savage/Fogarty Companies, Inc., a Dutch development firm, announced plans to build the TransPotomac Canal Center office complex on north waterfront land that included the Tidal Lock site. This property was involved in a title dispute with the federal government, which claimed ownership of all land along the Alexandria waterfront that had been created through filling since 1791. As part of the court-approved settlement, it was agreed that the lock and a portion of the basin would be excavated archaeologically and restored as the centerpieces of a public park. Hahn and Emory Kemp, a restoration consultant, were hired by the developer to excavate and restore the lock and part of the basin. This was successfully completed in 1987.

The canal had come full circle from its beginnings as a means for conducting trade between Alexandria and the hinterlands, to abandonment and burial, followed by rediscovery, and, finally, restoration with the hope of benefiting the public and drawing businesses in to fill the new offices.

Economic and Social Effects of the Alexandria Canal

The history of the Alexandria Canal, however, is much more than this. There were direct effects of building and operating the canal, but there were also indirect effects of this commercial venture. Alexandria's economy in the 19th century was revived by the establishment of the canal, which connected the town with the C&O at Georgetown. The hope that the C&O would be completed to Cumberland, Maryland, by the time the Alexandria Canal was completed was not fulfilled. Instead of coal, therefore, agricultural products became the major cargoes. This resulted in a flourishing trade, especially in the 1840s when the Irish famine greatly increased demand for American flour and grain. Alexandria merchants long referred to 1846, which brought so much misery to Ireland's poor, as that "glorious epoch" (Peterson 1932: 111). The Alexandria market quotations for flour delivered by canal in 1844 and 1845 were about 15 cents per barrel higher than "wagon prices," but the flour the canal brought from Frederick, Loudoun, and the surrounding counties was considered to be of superior quality to the local flour delivered by wagons (Hahn 1982: 89). Corn and lumber were also major canal cargoes and continued to be important even after coal became the real money-making commodity. The figures for Alexandria's ascending and descending canal traffic in 1847 were printed in the *Alexandria Gazette* (3 October 1847: 3).

In the 1840s, Baltimore offered serious competition to Alexandria as a market for agricultural products by way of the Baltimore and Ohio Railroad. The completion of the C&O Canal to Cumberland, Maryland, in October, 1850, made Alexandria an important coal terminus and tended to

Descending		Ascending	
Boats	151	Boats	140
Wheat	31,226 bu.	Fish	2,264 bbls.
I[ndian] corn	73,083 bu.	Salt	220 half-tons
Oats	4,355 bu.	Plaster	427¾ tons
Bran, shorts, etc.	11,492 bu.	Lumber, Shingles, etc.	213 half-tons
Whiskey	158 bbls.	Fresh fish and Sundries	
Corn meal	10,424 bbls.		

offset these losses (Peterson 1932: 111). The cost to ship coal dropped dramatically, from about 20 cents per ton mile in 1817 to one-quarter of a cent per ton mile in 1850 (Sharrer 1977: 28). Coal traffic on both the C&O and Alexandria canals steadily rose in the 1850s as America's industrialization continued. In 1857, canal boats transported about 177,619 tons of coal to Alexandria, accounting for about three-fifths of the total tonnage of coal carried by the C&O (Hahn 1982: 96). The ever-expanding use of the steam engine compounded the demand for coal. By 1860, four coal yards were operating on the Alexandria waterfront, and agents of the Cumberland, American, Borden Mining, and Frostburg coal companies had offices in town (Sharrer 1977: 28). A promotional piece for Alexandria carried on the Sanborn Fire Insurance Map in 1885 stated:

> The high price of coal in England has recently developed a new trade in the shipment, from Alexandria direct to Aspinwall [England] and San Francisco, of large supplies of Cumberland coal for the use of the steam lines engaged in the China and Japan seas; and in case these prices are maintained, it is reasonable to suppose that, in a short time, the ports of the West Indies, Central and South

FIGURE 3. The Alexandria Canal Tidal Lock and Basin and wharves on the Potomac River, ca. 1861-1865. *Courtesy of the National Archives, Washington, D.C.*

America will be supplied in a large degree, by shipment from Alexandria.

Products also were shipped *from* Alexandria by canal boat. Commission merchants sold goods made in Alexandria, such as wagons, agricultural tools, pottery, and tinware (Hahn 1984: i).

So-called "waste water," excess water diverted from a canal, was an energy source taken advantage of on the C&O Canal. Commercial opportunities through water-powered enterprises were anticipated for the Alexandria Canal as well. Such an opportunity was lauded in an 1838 *Alexandria Gazette* advertisement for land "intersected by the Alexandria Canal, which is elevated upwards of thirty feet above tide water, affording to the enterprising manufacturer situations for milling & manufacture purposes rarely to be met with ..." (*AG* 24 October 1838: 3). These possibilities were never realized, and the Alexandria Canal Company blamed Georgetown for excessive use of the water from the C&O, while the C&O Canal Company blamed the lack of water on leaks in the aqueduct. In 1857, one writer in the *Alexandria Gazette* asserted that anyone who built a cotton factory or flour mill which was dependent on the water from the Alexandria Canal was foolish. He went on to say it was obvious that the wooden trunk of the aqueduct would soon decay and fall into the Potomac, ending the arguments over water rights (*AG* 3 July 1857: 3). The idea that the canal could provide drinking water for Alexandria was entertained at the time when the Alexandria Canal Company was being formed. However, when the canal opened this plan was "quickly abandoned" (Wedderburn 1907: 13). The town would not receive water other than through wells and water carts until 1852, when the Alexandria Water Company began supplying water through its pipes.

One important stimulus to Alexandria business was through the services required by the canal boatmen. Shipyards repaired the vessels while merchants provided supplies for those living aboard. Boatmen gave business to taverns, restaurants, various shops, and houses of pleasure such as the "Do Drop In" (Miller 1988: 91). An article in an 1872 issue of the *Alexandria Gazette* mentioned the "large trade with the canal men, who nearly all purchase their supplies in our city. There are several hundred of this class of persons to be supplied with food and clothing and it amounts to a pretty large sum in the course of a year" (*AG* 6 December 1872: 3).

An unanticipated use of the Alexandria Canal aqueduct occurred on May 23, 1861. That night, three contingents of federal troops crossed the Potomac River to capture Alexandria, since Virginia had voted to join the Confederacy the previous day. A column of soldiers under a Major Wood crossed from Georgetown over the aqueduct but found the towpath constricting. Early in the winter of 1861-1862, the aqueduct was drained of water, and the trough was planked over for use as a double-track wagon road (Mitchell 1981: 5). Throughout the Civil War the "Aqueduct Bridge" served as a convenient conduit for troops and supplies. The wharves at the outlet of the Tidal Lock (Figure 3) were in nearly constant use by the U.S. Army Quartermaster for the unloading and shipment of both troops and supplies.

The Alexandria Canal was in a state of ruin following the four years of abandonment during the war. The *Alexandria Gazette* carried the following description:

> ... owing to long disuse and exposure to sunshine and wind the banks of the canal have become cracked and the aqueduct decayed, the coal boats have rotted or been used for other purposes, the vessels have long since gone from the wharves, the coal agents have moved their depots to other cities and an occasional cart, hauling a load of Philadelphia coal, is the only evidence of life about the once active coal yards. The debt of the Canal is so large that but little benefit is expected ever to accrue to the stockholders, but the city can be immediately affected by a resumption of trade on this work (*AG* 12 July 1865: 3).

The canal was eventually brought back into operation and was used almost exclusively for coal transport, which peaked about 1877 and then declined throughout the 1880s until the closing of the canal in 1886. Even at this time there was some economic benefit from the canal. The ice that formed in the 500 x 50 foot turning basin (see Figure 2) was cut commercially and profitably during the winter months for nearly 50 years (Tilp 1978: 231). In the 1890s, stones from the walls of the canal locks were removed by the city and used as road-building material (*AG* 11 March 1898: 3).

The impact of the canal on Alexandria's society is less easily explained as so many of the canal records discuss only economic concerns. Nevertheless, there were and are effects on the urban community. Active trade and healthy profits obviously touched a large segment of the population from tavern owners and merchants, to blacksmiths and even police.

The citizens of Alexandria felt the effects of the canal even as it was being constructed. In 1838, an advertisement appeared in the *Alexandria Gazette* for "one hundred laborers wanted on the Alexandria Canal, near the Georgetown Ferry to whom constant employment and very liberal wages will be given" (*AG* 16 August 1838: 3). Many of the Welsh, English, and Irish workers who were recruited in Britain to work on the C&O Canal arrived at the port of Alexandria. They were indentured to work for four months on the canal in return for their passage. Some broke their indenture, many ended up in jails or in alms houses, some undoubtedly settled in the area (Kytle 1983: 32-33). Many of these laborers must have helped build the Alexandria Canal.

An African-American community grew up around the locks at the north end of Alexandria. It was called "Cross Canal," and it is likely that many of its residents found employment in jobs associated with the Alexandria Canal. Mrs. Virginia Knapper, a long-time resident of the neighborhood, remembered that as a child around the turn of the century "the canal was quite deep ... when it was freezin' we used to skate on it. We had no skates, [we used] our feet ... used to do it across from the school ... used to be late 'cause we were having fun" (Knapper n.d.: 2). Skating on the canal must have been a winter activity enjoyed by Alexandrians for many years. A newspaper article in January 1868 reported that "the skating carnival of the past three to four days was enjoyed to its fullest extent by those fond of the sport and the canal basin was the scene during the continuance of much youthful recreation" (*AG* 14 January 1868: 3). There is an account as well of the basin being the site of religious services held on a canal boat for the benefit of the boatmen docked in this body of water (*AG* 29 April 1856: 3).

Some members of Alexandria's African-American community also conducted a clandestine trade with the canal boatmen. An unidentified C&O Canal boathand, in an account of a trip in 1859, described how his captain watched for signals at night from what he called "Montgomery County merchants" (Clark 1960: 117). Upon hearing a signal, the captain brought his boat to the bank to meet the trader. An equality rooted in economics prevailed in such meetings. No one asked if the man was free or slave, and no questions were asked as to how the commodity—be it a sack of flour or a pig—had been obtained. The author remarked how there was "little risk in the business for neither the slaves or the poor whites had any interest in stopping it" (Clark 1960: 117).

Archaeological Excavation and Restoration of the Lock and Basin

While the role of the Alexandria Canal in the 20th century has not been as profound as that in the 19th century, it is still the source of social and economic benefits. In the 1970s, interest in historic preservation and planned waterfront development provided the impetus for restoring the Tidal Lock in a park setting. In 1974, Thomas Hahn, president of the American Canal Society, expressed his opinion in the *Washington Post* that "the Alexandria Tide Lock clearly has the potential of becoming one of the most memorable and valuable features of the Alexandria waterfront" (*WP* 6 July 1974: A15). To increase public awareness of the Tidal Lock site, a small exhibition on the canal was set up in 1977, and a painting of the historic lock was presented to the mayor of Alexandria at the hotel overlooking the site (Digilio 1977: 13). Citizens and city officials began to recognize the historical significance of this waterfront property.

The process of locating and preserving the Tidal Lock began in 1978. Terry Klein, then archaeologist with the Alexandria Regional Preservation Office attached to Alexandria Archaeology, led a team of volunteers in excavating test trenches near the foot of Montgomery Street. The tops of the stone walls of both the lock and basin were uncovered. This confirmation of the location and preservation of the lock 2-4 feet below the ground surface allowed the site to be entered on the National Register of Historic Places in 1980.

The Alexandria Archaeological Commission, the citizen advisory group for Alexandria Archaeology, and city administrators supported the effort to preserve the lock. The city's Department of Planning and Community Development and Alexandria Archaeology applied for and received a grant from the National Trust for Historic Preservation for the purpose of "protecting, maintaining and interpreting the only restorable archaeological property on the waterfront" (Harman 1981: 1). In March 1982, Thomas Hahn directed the excavation of the site, progressing from the river end of the lock westward, to uncover the upper walls of the entire lock chamber and the eastern portion of the adjoining basin. The goal was to assess the condition of the lock and to determine what would be necessary to restore it. The lock measured 90 feet long, 16 feet 8 inches wide, and 15 feet 6 inches deep. A small section of fill was removed to examine the walls and floor of the lock chamber. The walls were composed of cut blocks of "aquia freestone" (local sandstone) mortared with natural hydraulic cement. The floor was lined with wood planking. The structure was in excellent condition with the exception that some of the coping (the top or capping layer of stones) was missing, presumably having been robbed for building material. A large portion of the basin walls, which were dry-laid stone, had been removed in the past as well. The basin measured about 80 feet in width, about 268 feet in length, and approximately 8 feet in depth (Hahn 1984: 77, 91).

The remains of a lock gate were found at the western end of the lock chamber adjacent to an in-place heel post from which the gate had hung. An iron wicket gate and two wicket gate frames were found in addition to pieces of the large gate timbers. Wicket gates were iron plates that pivoted on a vertical axis so as to regulate the flow of water through the closed lock gates.

The fill in the lock and basin consisted of soil and various types of debris that had been dumped into these abandoned troughs in the late 19th and early 20th centuries. A large portion of the fill was partially melted glass vessels and other manufacturing by-products from the Old Dominion Glass Factory located on the adjacent blocks to the west. This company produced various types of glassware from circa 1898 to 1927; it found the old canal basin and lock convenient repositories for manufacturing waste.

FIGURE 4. The Alexandria Canal Tidal Lock (looking west) in June 1986, showing gravel being placed around the lock gates in the foreground. *Courtesy of Alexandria Archaeology, Alexandria, Virginia.*

Evidence of the canal towpath was not discovered, but identification of this feature had not been a primary objective of the excavation project. It is also possible that no towpath was present at the locks, or at least at the Tidal Lock, and that boats were pulled by hand or poled through the lock chamber (Hahn 1984: 86-87).

After the 1982 investigation of the canal was completed, both the lock and basin were reburied until 1984 when their excavation was prompted by plans to build a large office complex, the TransPotomac Canal Center. The developer, Savage/Fogarty, agreed to restore the Tidal Lock in a park setting as part of the property title settlement. Although the restoration was a condition of the purchase of this land, the lock with a surrounding park was viewed as a positive public relations feature as well. This site would act as a focus of historic interest within the office complex and serve as an added amenity for the firms occupying the nearby offices. Space in a building adjacent to the lock was also provided for the creation of a city waterfront museum.

Savage/Fogarty contracted with Thomas Hahn and Emory Kemp to direct the excavation and restoration of the lock and part of the basin. In 1984 and 1985, the Tidal Lock and the eastern end of the basin were completely excavated. The significance of this site was increased when the remains of three lock gates and their associated iron fittings were discovered in the lock, below the concrete slab of a 20th-century structure. The top one-third of each gate had been cut off during the construction of a shipyard boat shop built circa 1917 (Tilp 1978: 82), but the lower portions of the gates were in excellent condition (Kemp and Hahn 1985: 1). Each gate contained two iron wicket gates, paired at the bottom on the heel-post side of the gate. Although no exact date could be established for the gates (lock gates had an average service life of 12 to 15 years), it was the opinion of Kemp and Hahn (1985: 1) that they probably dated to circa 1865-1870, when the canal was restored after the Civil War. Furthermore, they believed that no other complete lock gates have survived from this period of American history. [See the article by Carlisle et al. in this volume for another archaeological example of a canal gate—Ed.]

Once found, the canal gates presented a difficult preservation challenge. Conservation would be a very expensive and labor-intensive undertaking for the City of Alexandria. After evaluation of alternatives and consultation with conservators and members of the city's public commissions, the gates were reburied *in situ*.

It had been part of the restoration plan for the lock to lay a concrete slab 4-5 feet below the top of the wall for engineering and public safety reasons. The gates were placed upright in the bottom of the lock and gravel was filled in around them for support (Figure 4). A thick layer of concrete was poured above the gravel and painted a dark gray so that the water in the lock and basin would appear to be of greater depth. The natural water table percolates through the gravel under the concrete slab to create an anaerobic environment for the gates.

The missing coping stones of the lock were replaced with concrete, and the missing stones in the basin walls were replaced as well. The lock was repointed with hydraulic cement similar to the original. Historically accurate timber gates with iron and steel fittings were fabricated and installed. A wooden bridge was constructed over the east end of the lock to duplicate the bridge shown in the Civil War period photograph of the Tidal Lock (see Figure 3). Tidal flow from the Potomac River was controlled by the placement of a concrete barrier wall across the outlet of the lock.

Conclusion

With the completion of the restoration of the Tidal Lock in 1987, this artifact of the engineer's art had been transformed from its commercial role in the 19th century to an interpretive role as the centerpiece of an historical park. In its new role, the lock still retains an economic function by providing an historic ambience that attracts businesses to purchase space in the surrounding office buildings. Tourists and city residents alike are drawn to the waterfront. Its main contemporary role in the community, however, is cultural in serving as a tangible link to Alexandria's past. Alexandria has long been known as a city proud of its heritage. The restored lock and the exhibitions in the adjacent Waterfront Museum are reminders of that heritage—of a grand commercial venture undertaken by Alexandria's businessmen and the many lives that were affected as a result of this endeavor.

The editor of the *Alexandria Gazette*, James W. Coldsmith (1985), expressed the sentiments of many historically minded Alexandrians about the restoration of the Tidal Lock and at the same time recognized the closing of another phase in the canal's history. He wrote:

> It is a fitting and happy circumstance that this commercial relic of the past was restored to us by a modern business firm for a solid commercial reason. The restored lock will become the interest center for Savage-Fogarty's new TransPotomac Canal Center development.
>
> That's why the company was willing to spend the money needed to excavate the lock, and it is a reason the businessmen who spent the money to build it 150 years ago would understand completely (*AG* 4 October 1985: A4).

References

Alexandria Gazette [Alexandria] (*AG*)

1831 6 July 1831: 3.
1838 16 August 1838: 3, 24 October 1838: 3.
1839 29 April 1839: 2.
1843 4 December 1843: 2.
1847 3 October 1847: 3.
1856 29 April 1856: 3.
1857 3 July 1857: 3.
1865 12 July 1865: 3.
1868 14 January 1868: 3.
1872 6 December 1872: 3.
1894 11 May 1894: 3.
1898 11 March 1898: 3.
1985 4 October 1985: A4.

Clark, Ella E., editor

1960 "Life on the C. & O. Canal: 1859." *Maryland Historical Magazine* 55: 2.

Coldsmith, James W.

1985 "A New One." *Alexandria Gazette* 4 October 1985: A-4.

Corporation of Alexandria

1836 *Memorial from the Corporation of Alexandria to the Honorable the Senate and House of Representatives of the United States of America in Congress, December 27, 1836*. Document on file, Alexandria Archaeology, Alexandria, Virginia.

Department of Planning and Community Development and Alexandria Archaeology, Office of Historic Alexandria

n.d. *Alexandria Canal.* Brochure prepared by the Department of Planning and Community Development and Alexandria Archaeology, Office of Historic Alexandria. On file, Alexandria Archaeology, Alexandria, Virginia.

Digilio, Alice

1977 "Remembering the Alexandria Canal." *Washington Post* 19 May 1977: Virginia Section, 13.

Hahn, Thomas

1974 "Alexandria Canal Lock." *Washington Post* 6 July 1974: A15.

1982 "Maritime Historical Report for Alexandria Canal Tide-Lock Project." Ms. on file, Alexandria Archaeology, Alexandria, Virginia.

1984 "The Industrial Archaeology of the Tide Lock and Pool No. 1 of the Alexandria Canal: Revised Draft." Ms. on file, Alexandria Archaeology, Alexandria, Virginia.

Harman, Douglas

1981 *Memorandum to the Alexandria City Council*, 6 February 1981. On file, Alexandria Archaeology, Alexandria, Virginia.

Hopkins, G.M., and Company

1877 *City Atlas of Alexandria, Va. from Official Records, Private Plans and Actual Surveys Based Upon Plans Deposited in the Department of Surveys*. Philadelphia, Pennsylvania: G.M. Hopkins and Company. Library of Congress, Record G1294.A3H6 1877, Washington, D.C.

Kemp, Emory, and Thomas Hahn

1985 *Memorandum Concerning the Historical Significance of the Lock Gates Uncovered in Lock Chamber Excavation*. On file, Alexandria Archaeology, Alexandria, Virginia.

Knapper, Virginia

n.d. *Alexandria Reflections, Life Around the Canal*. Leaflet on file, Alexandria Archaeology, Alexandria, Virginia.

Kytle, Elizabeth

1983 *Home on the Canal*. Cabin John, Maryland: Seven Locks Press.

Miller, T. Michael

1988 *Murder and Mayhem, Criminal Conduct in Old Alexandria, Virginia, 1719-1900*. Bowie, Maryland: Heritage Books Inc.

Mitchell, Vivienne

1981 "The Alexandria Canal." *American Canals* 38: 4-5.

Peterson, Arthur G.

1932 "The Alexandria Market Prior to the Civil War." *William and Mary Quarterly*, Second Series, 12: 104-114.

Sanborn Map and Publishing Company

1885 Sanborn Fire Insurance Map of Alexandria, Virginia. New York: Sanborn Map and Publishing Co.

Sharrer, G. Terry

1977 "Commerce and Industry." In *Alexandria: A Town in Transition, 1800-1900*, J.D. Macoll and G.J. Stansfield, editors. Alexandria, Virginia: Alexandria Bicentennial Commission and Alexandria Historical Society: 16-38.

Tilp, Frederick

1978 *This Was Potomac River*. Alexandria, Virginia: Frederick Tilp.

U.S. Congress

1841 Map published in 1841 from surveys by Lt. Col. Kearney, Major Trumbull, W.M.C. Fairfax, and M.C. Ewing. In *Report of the War* ... 26th Congress, Second Session, Senate Document 178 (Serial Set No. 378, 5 February 1841). Washington, D.C.: U.S. Government Printing Office.

Washington Post [Washington, D.C.] (*WP*)

1974 6 July 1974: A15.
1977 19 May 1977: Virginia Section, 13.

Wedderburn, Alex J.

1907 *Souvenir Virginia Tercentennial of Historic Alexandria, Virginia*. Washington, D.C.: Sudworth Printing Company.

Acknowledgments

The excavation and restoration of the Alexandria Canal Tidal Lock, which made the writing of this article possible, involved the efforts of many people. Thomas Hahn was instrumental in reminding the city of the canal; Ellen Pickering and Vivienne Mitchell took up the cause and raised support for preserving the Tidal Lock site. The Alexandria Archaeological Commission, with Bernard Brenman as chairman, played a consistent and active role in seeing that the preservation work was completed. George Colyer and John Stephens produced the grant application to the National Trust for Historic Preservation that funded the major exploratory excavations at the lock and basin. Thomas Hahn and Emory Kemp served diligently as the experts who saw that the preservation measures were carried out properly. Savage/Fogarty Companies, Inc., provided the funds to save this historic feature of the Alexandria waterfront for future generations. City of Alexandria staff worked with the developers to accomplish this goal. Thanks go to all these individuals and organizations and many others who lent their support and time as interested citizens and volunteers.

RICHMOND'S GREAT BASIN EXCAVATION AND ITS CONSEQUENCES

William E. Trout, III
A. Howe Todd

Abstract

From 1800 to 1880, dozens of James River Batteaux and canal boats sank accidentally or of old age in the "Great Basin" in downtown Richmond, Virginia, the eastern terminus of the James River & Kanawha Company's canal and turnpike system from the Ohio River to the Atlantic seaboard. Salvage archaeology in the basin during high-rise building in 1983-1985 revealed construction details of the James River Batteau, a patented craft invented in 1771 to carry tobacco hogsheads and farm goods down the James and other white-water rivers. This discovery inspired the construction and pioneering voyage of a batteau replica, which led to the annual James River Batteau Festival, a 140-mile voyage now involving over two dozen batteaux, several hundred participants, and thousands of spectators. This has brought about a new interest in Virginia's river history, in re-enacting the boatman's life, and in the potential of rivers for recreation and tourism. It has also accelerated interest in historic canal preservation and parks, especially the restoration of the Kanawha Canal through downtown Richmond as an historical amenity and tourist attraction to revitalize the stagnating inner city.

Introduction

The year 1991 marked the bicentennial of George Washington's visit to Richmond, Virginia, and his ceremonial voyage up the James River Canal

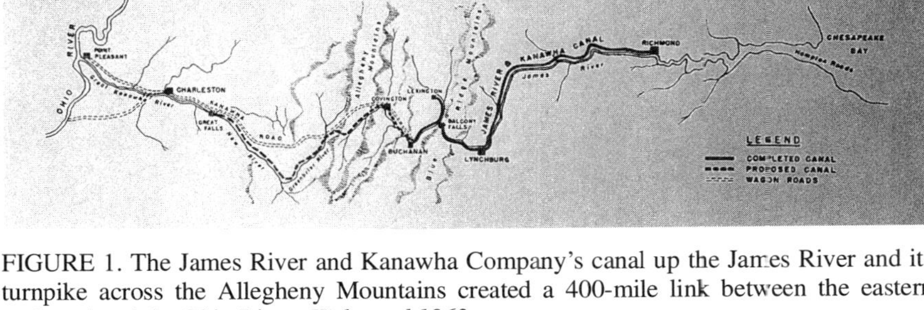

FIGURE 1. The James River and Kanawha Company's canal up the James River and its turnpike across the Allegheny Mountains created a 400-mile link between the eastern seaboard and the Ohio River. *Kirkwood 1963.*

(Figure 1) in 1791 (Trout 1992). Washington was a major supporter of the canal, and was honorary president of the canal company. On April 12, 1791, he traveled up the canal, through the locks, and through the great arch at the entrance, which is still intact in Three-Mile Locks Canal Park (Figure 2). Washington and his entourage traveled on two wooden white-water boats known as James River Batteaux, which had been invented 20 years earlier to carry tobacco hogsheads down the James to Richmond (Figures 3, 4).

The canal was completed into downtown Richmond in 1800, when the Great Basin was finished and filled with water (see Figure 2). This basin was the Richmond terminus of the canal, a large body of water three blocks long and a block wide. Batteaux from as far away as the Allegheny Mountains, 250 miles upriver, navigated down the James and its branches, loaded with hogsheads of tobacco, with flour, pig iron, coal, and other products of Virginia's farms and mines. They came to unload in the Great Basin, the commercial center of Richmond, three blocks straight down the hill from the state capitol. Some of the boats were sold for lumber when they arrived, others returned upriver loaded with city goods. It is also now known that a lot of boats, perhaps 200 of them, stayed in the Great Basin where some of them sank accidentally and other boats, too old for another voyage, were abandoned (Trout 1986).

Over the years from 1800 to 1880, the Great Basin gradually silted up halfway with mud. From time to time, other boats sank and settled in the mud, the oldest, the batteaux, at the bottom. In later years, when larger canal boats came into use, they sank into the mud closer to the top. Now

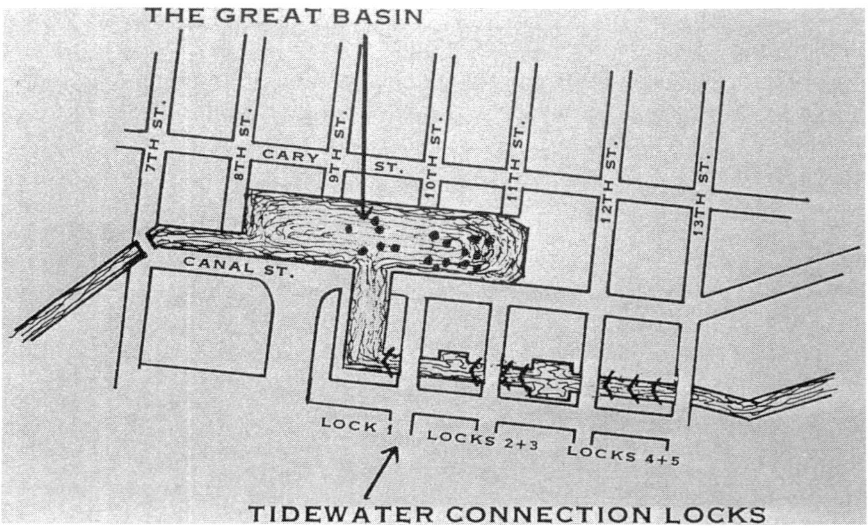

FIGURE 2. The Great Basin and Tidewater Connection Locks in downtown Richmond, Virginia. The dots indicate the locations of some of the batteaux identified in the Great Basin during the construction of office buildings in 1983-1985. *Virginia Canals and Navigations Society 1985.*

and then a cargo was lost—a load of grindstones, a bale of leather, some bars of pig iron. The basin was extraordinarily deep in the middle, up to 40 feet in places, so sunken boats were left undisturbed in the depths, and the water was too deep and murky to risk a life diving for lost cargo. For 80 years everything piled up, left where it was. Also piling up in the mud was garbage and trash — broken pottery, empty bottles, old shoes, broken tools, chicken bones, and anything else the boatmen or the citizens of Richmond wanted to be rid of. The result of all this was a stratified urban archaeological site covering the basin's 80-year life span from 1800 to 1880. There are still parts of this stratified site left intact, under the streets, waiting for future archaeologists with big grants and a lot of time.

Excavation of the Great Basin

In 1880, navigation on the canal stopped when what is now the C&O Railway, or CSX, took over the canal, put their tracks on the towpath, and made the Great Basin into a rail yard. Later, the land became prime real estate in downtown Richmond, and in 1983 CSX began to build high-rise buildings in the basin. Even then it was a recognized historic site that had been listed on the National Register of Historic Places, but Richmond had no city

FIGURE 3. One of a pair of stereo views of the "Red Boat," the only painted batteau discovered in Richmond's Great Basin. Found at the bottom of the basin silt and held together with wrought nails, it is one of the oldest batteaux to be excavated. During the project, stereo views were used to generate scale drawings of several boats found in the Great Basin. *Photo: Virginia Department of Transportation, 1985.*

FIGURE 4. Photo-mosaic of the excavation of a metal-hulled packet (passenger) canal boat in Richmond's Great Basin on July 2, 1985, by students from Virginia Commonwealth University. The boat may have been the *Joseph C. Cabell. Photo: George Davis Rawls.*

FIGURE 5. Artist's impression of a dramatic moment in the life of a batteau crew that appeared in *Harper's Weekly* on February 21, 1874.

archaeologist, and had no say over the fate of this privately owned archaeological site. Before construction work in the basin began, Dr. L. Daniel Mouer of Virginia Commonwealth University tried to fund an archaeological study of the site, but he was unsuccessful. So it was up to canal "buffs" to keep an eye on the excavation. Jimmy Moore and Bill Trout kept watching through the fence as the backhoes dug, but it was not until some friendly workmen let them go down into the hole that anyone realized there were boats down there. First the newspapers were called, then archaeologist Lyle Browning. The developer, Henry Faison of North Carolina, was very cooperative, agreeing to allow time to study and remove the first boats and then consenting to the monitoring of the rest of the excavation, studying more boats, and saving what could be saved (Trout 1983).

The project was a joint one between the Archeological Society of Virginia and the Virginia Canals and Navigations Society. Altogether, the locations of nearly 60 boats, most of them batteaux, were mapped, and selected parts of a dozen of them were saved. The wood from these boats is underwater now, waiting to be treated and reassembled. The other artifacts from the basin are also waiting to be studied to reveal their stories about life in and around the Great Basin during the canal era.

FIGURE 6. The tremendous publicity surrounding the excavations in Richmond's Great Basin has renewed interest in canal boats and river travel. Here, the crew of the *J.R. Mud Cat* contend with the rapids of the James River in Fluvanna County, Virginia, during the annual River Batteau Festival. *Photo: Overton McGehee,* Richmond Times-Dispatch, *June 24, 1989.*

A Second "Batteau Era"

Back in 1983, the excavation of the boats created quite a stir in downtown Richmond. Many people say they spent a lot of time looking out at the project from their office windows instead of working. But what happened after the excavations were completed was totally unexpected and amazing. As a result of the excavation of batteaux, Joe Ayers, a folk musician, re-enactor, and historian who lived near the river in Fluvanna County, Virginia, built a replica of one of the canal boats, navigated it down the James River, and got others so excited and involved that there is now an annual James River Batteau Festival (Figures 5, 6). By 1991 over two dozen batteaux were in operation. Their crews take a week off every year from their regular jobs to participate in the festival. Some 200 costumed crew members navigate these boats down the white-water of the James River for 140 miles, stopping at local festivals every night at the river towns. Thousands of people visit the boats and the boatmen's camps, buy batteau T-shirts and scale models, and learn about the river and about a period in Virginia's history that in 1983 was of special interest to only three or four people.

During the rest of the year one can see batteaux on other rivers in and around Virginia, re-creating similar scenes to those one might have come upon two centuries ago. James River Batteaux were featured in "National Geographic Explorer," a 1988 national television special; they've been on the cover of *National Geographic*; they've even been written into a science fiction novel (Rucker 1990). Artists have painted them; there's a batteau sculpture in Richmond, a coffee-table book called *Batteaux*, and there's a periodical called the *Virginia Batteau Journal*. The Virginia state legislature designated the James River from Lynchburg to Richmond, Virginia, as the James River Batteau Festival Trail, and Lynchburg named its farmers' market the "Batteau Landing" (Trout 1988).

These and other consequences of the work in the canal's Great Basin make history exciting, but for many people it is the direct physical experience with these batteaux and the experience of using them on water that are most important and that evoke a personal connection with the past. As a result, one can also see batteaux displayed on long trailers in Christmas parades and in county festivals, in parks for Fourth of July celebrations, or out in the water for events called "Batteau Day" and "Canal Day." Batteaux also make an appearance at the annual Rucker family reunion. It was two Rucker brothers, Anthony and Benjamin, who invented the James River Batteau in 1771, and the family later patented it. The batteau has become the centerpiece of the Rucker family reunions, and the whole clan keeps an eye out for new information about these vessels.

All of this renewed local interest in canal vessels, the revival of the skills needed to navigate them, and the heightened enjoyment of the recreational aspects of canals, water, and rivers in contemporary urban life, started with the Great Basin excavation in Richmond. In many ways, Virginia has entered a second "Batteau Era." It's a James River Batteau revival, and lots of people are having lots of fun and learning a good deal in the process because of what was excavated and saved in 1983. Perhaps this interest in batteaux will spread to other states and will last a long time. One can only imagine that 50 years from now, there may be "old" batteau crew reunions, where the "old-timers" will line up to have their pictures taken and to reminisce about how different it was back in the "classical" period of the 1980s and 1990s.

Such speculations are a lot of fun, but of course there is a more serious side to all of this. Those same people who are having a good time building and poling batteaux and camping out on the riverbank, and the thousands of people who have visited the festivals, also have been introduced to the river and to the batteau era. It has given many James River towns something to be proud of again. The experience has also afforded many people a new perspective on history, a window into a period two centuries ago when rivers and canals were the nation's superhighways and batteaux and canal boats were modern transportation technology. To believe that an appreciation of one's place in history is good for society and contributes to its stability and to quality of life is also to believe that the "batteau revival" in its full context has done a lot to instill a deeper appreciation of archaeology's potential to contribute to the meaning of late 20th century American life.

FIGURE 7. Reynold's Metals Company's Tidewater Canal Locks Park in downtown Richmond, Virginia, looking east from the locks toward the stone arch of the 13th Street Bridge. *Photo: W. E. Trout, III, 1984.*

The people who are out on the river have also, of necessity, rediscovered the lost art of poling and navigating a batteau. This is a kind of knowledge that extends directly from the experience of material culture. Some of these people are also rediscovering and mapping the channels and wing dams built for batteaux two centuries ago. In the process, they are accumulating a body of experiences and adventures worth recording as oral history and as experimental archaeology, because the experiences they have to relate may be analogues to those from the original batteau era.

The resurgence of interest in the James River & Kanawha Canal and in the James River Batteau has also prompted new academic research, for example that by Bruce Terrell (1988) and David Brown (1988). Such research helps to construct a framework for future investigations on this kind of canal craft, its origins, and on the people who used it.

In considering the future of this city's canal, it is important to point out that the idea of restoring the canal in downtown Richmond has been around for a long time. At each end of town, the canal still carries water, but it's the mile in the middle, through downtown Richmond, that poses the greatest problem but also the greatest potential. This is the part of the city that will benefit most from a renewed acquaintance with both water and boats, the creation of an urban waterscape to revitalize the inner city for Richmond's citizens and tourists alike. The slogan that has been coined to express this idea is "Bring the Water to the City" and in the process help downtown pay for itself. The idea is to adaptively reuse as much of the old canal as possible, to join the parts of the canal back together again and to bring it back to life.

The effort to restore the canal has been led by one of us (A. Howe Todd), who for many years was a Richmond city planner, then acting city manager, and who has now retired with the goal of seeing the canal at last restored. It has been an uphill battle, dealing with the city bureaucracy and with a rapidly changing cityscape. The canal basin is now full of buildings, and three of its stone locks are gone. But two locks remain in a park, thanks to Reynolds Metals Company (Figure 7), and there is a corridor, still undeveloped, where a new canal can be dug to connect the parts together again. At the moment, the major concern is Richmond's huge floodwall, which is now under construction. It crosses the path of the canal in two places, so it is vital to have gates in the wall that can be used by the canal and the boats.

Readers are invited to visit Richmond's canal. One can drive to the Great Ship Lock, the Reynolds Metals exhibits at the Tidewater Connection Locks, and Three-Mile Locks and George Washington's arch at Byrd Park. All of these are in urban riverside parks. Tourists can also stroll along the canal downtown by following the historical signs that accompany Richmond's "Canal Walk," a two-mile route that encompasses the Great Basin, the locks and exhibits at Reynolds Metals, a canal basin artifact exhibit, and a batteau sculpture on Brown's Island. Ironically, one cannot yet see a boat on the canal, but the time is coming when Richmond's canal will be once again a vital part of the city, and when everyone, even the city government, will wonder how they ever got along without it.

References

Brown, David

1988 *Batteaux: That Old Time Boating on the James*. Charlottesville, Virginia: Arrowsmith and Lewis.

Garrett, Wilbur E.

1987 "George Washington's Patowmack Canal, Waterway that Led to the Constitution." *National Geographic* 171(6): 716-753.

Harper's Weekly

1874 "Running the Rapids of the New River in Virginia." *Harper's Weekly*, February 21, 1874.

Kirkwood, James J.

1963 *Waterway to the West*. Washington, D.C.: U.S. Department of the Interior, National Park Service.

Rucker, Rudy

1990 *The Hollow Earth*. New York: William Morrow and Company.

Terrell, Bruce G.

1988 *The James River Bateau: Tobacco Transport in Upland Virginia, 1745-1840*. Unpublished M.A. Thesis, Department of History, East Carolina University.

Trout, W.E., III

1983 "VC&NS Unearths Canalboats." *The Tiller* 4(2): 1-4. [Virginia Canals and Navigations Society, Rt. 2, Box 254, Lexington, Virginia 24450].

1986 *A Guide to the Works of the James River & Kanawha Company, from the City of Richmond to the Ohio River*. Lexington, Virginia: Virginia Canals and Navigations Society.

1988 *The James River Batteau Festival Trail*. Lexington, Virginia: Virginia Canals and Navigations Society.

1992 "Celebrating the Bicentennial of George Washington's Trip Up the James River Canal." *The Tiller* 12(4): 4. Lexington, Virginia: Virginia Canals and Navigations Society.

Virginia Canals and Navigations Society

1985 *Touring Richmond's Canal*. Virginia Canals and Navigations Society, Rt. 2, Box 254, Lexington, Virginia 24450.

REHABILITATION OF HISTORIC STRUCTURES ON THE DELAWARE & RARITAN CANAL IN CENTRAL NEW JERSEY

Brian H. Morrell

Abstract

Affiliated with Historic Conservation & Interpretation, Inc. of Newton, New Jersey, the author has been providing cultural resources inspection and consultation services for the past 10 years during maintenance and rehabilitation projects on the historic Delaware & Raritan Canal in central New Jersey. The D&R Canal, once an important link in the inter-regional anthracite trade, operated between 1834 and 1934 before its ownership passed to the State of New Jersey. Today, the canal serves three major functions: It is a conveyance for potable water, a state park, and an historic district. Most of the canal survives, although it has had little maintenance for more than 50 years, and it has undergone adaptation for water supply use. Recently, rehabilitation of the historic structures on the canal has begun, affording an opportunity to document the canal's construction. This article focuses on archaeological discoveries made during these maintenance and rehabilitation projects and discusses the philosophy behind the rehabilitation approach.

Introduction

In association with the cultural resource management firm Historic Conservation and Interpretation, Inc. of Newton, New Jersey, the author has served as a cultural resource/historic preservation consultant during the past 10 years on over two dozen projects on the historic Delaware & Raritan Canal (D&R Canal), which is located in the central part of New Jersey. Discoveries related to the canal's construction, made during a three-year dredging project, are discussed first followed by the results of various projects that have rehabilitated several historic canal structures. Finally, the rehabilitation philosophy developed during the course of the various projects is presented to facilitate the process for others who may be faced with preservation decisions for historic canals.

History of the D&R Canal

The D&R Canal's long and complex history is summarized in William J. McKelvey's *The Delaware & Raritan Canal, A Pictorial History*

(1975), which is the source of the following historical background on the canal.

The completion of the D&R Canal across the narrow central "waist" of New Jersey in 1834 was the fulfillment of a more than 100-year-old dream of creating an inland navigation system that would eliminate travel around the extreme southern end of New Jersey for passage between the Hudson and Delaware rivers (Figures 1, 2). Constructed between 1830 and 1834 by Irish immigrant labor, the waterway consisted of a 42-mile Main Canal that was 75 feet wide and 8 feet deep, and connected Bordentown on the Delaware River with New Brunswick on the Raritan River. The canal was an important navigation link between New York City and Philadelphia. It utilized 14 locks to climb 52.5 feet from the western terminus on the Delaware River to the canal's summit at Kingston, and then descend 55.8 feet from Kingston to the eastern terminus on the Raritan River. The Main Canal accommodated various types of vessels ranging from the mule-drawn canal boats of the region's other canals to steam-powered tugs and yachts. To protect the canal's banks from erosion caused by the wake of passing boats, continuous embankment stone, or riprap, was installed in 1853 along both sides of the Main Canal.

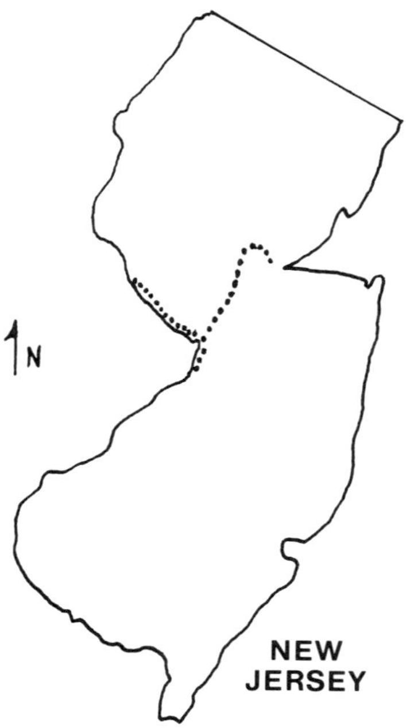

FIGURE 1. Map showing the location of the D&R Canal (dotted line) in the State of New Jersey. *Courtesy of the Delaware and Raritan Canal Commission.*

The Main Canal's primary source of water was a 22.5-mile Feeder Canal that began on the Delaware River at Bull's Island in Hunterdon County, New Jersey, and paralleled the river to Trenton where it joined the Main Canal. The Feeder Canal was 60 feet wide and 6 feet deep, had three locks, and was navigable for its entire length. Due to its smaller size, only mule-towed canal boats were utilized. Boats and log rafts could enter it at Bull's Island from the Delaware River; however, most of the traffic on the Feeder Canal entered at a lock constructed in 1847 at Lambertville, New Jersey, that enabled passage across the Delaware River at New Hope, Pennsylvania, from the Lehigh Navigation-Delaware Canal system.

A pioneer in the use of steam-powered vessels, the D&R Canal was primarily a link in a network of canals

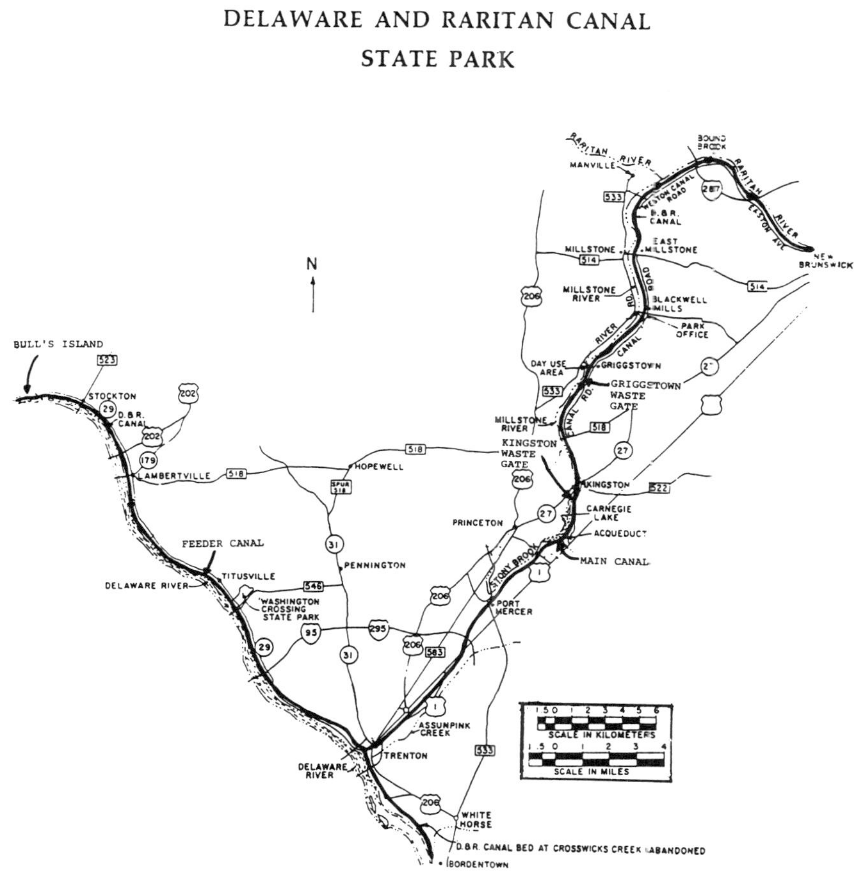

FIGURE 2. Enlargement map of the Delaware & Raritan Canal State Park. *Courtesy of the Delaware and Raritan Canal Commission.*

that extended from the eastern Pennsylvania coalfields to the New York metropolitan area. A peak year for the canal was 1866, when net earnings were $933,643 and 2,857,233 tons of cargo were shipped, 83 percent of which was coal.

In 1871, the D&R Canal was leased to the Pennsylvania Railroad Company for 999 years. Competition from new through-railroad links and unfavorable toll rates resulted in a continual decline in use of the canal until its abandonment in 1933. The State of New Jersey assumed ownership of the canal in 1934, and in 1944 began conversion of the waterway to a public potable water supply system. In 1973, the D&R Canal was entered on the National and New Jersey Registers of Historic Places and one year later became the Delaware & Raritan Canal State Park. Currently, in addition to serving as a water supply system, the canal is one of the state's major historic districts and is a 60-mile recreational corridor of ever-increasing importance in central New Jersey.

In 1974, the Delaware and Raritan Canal Commission was created by the New Jersey state legislature to protect the canal park's recreational and historic aspects and water quality. The water supply function of the canal is under the jurisdiction of the New Jersey Water Supply Authority (NJWSA), which leases the canal and operates it to sell water to municipalities, industries, and private concerns.

Curiously, there are few known surviving corporate records from the D&R Canal's managing company; the currently available engineering documentation relating to its physical structure is almost completely non-existent. Although no exhaustive region-wide search for the waterway's records has been undertaken, New Jersey repositories are nearly devoid of D&R Canal information, and even the New Jersey State Archives' recent acquisition of part of the Pennsylvania Railroad's archival holdings has produced no engineering information about the canal. Unquestionably, a thorough regional search for the canal's engineering records is warranted and should be undertaken to enable a logical and complete cultural resource management plan to be formulated.

In contrast to the lack of historic documentation, the majority of the physical canal prism and its many varied 19th-century structures survive mostly intact, although greatly neglected. As a result, the remaining architectural and historic fabric of the waterway itself is all the more significant, not only because it has survived, but also since it is the primary source of information on the canal's historic engineering design.

No comprehensive assessment of the Delaware & Raritan Canal's economic impacts on Philadelphia and New York can be made. This is due to the scarcity of the canal company's surviving records and to the absence of a substantive economic study of the impact of the canal's operation on either city or the intervening region. Nevertheless, a few preliminary assertions can be offered.

First, the canal company was part of a joint company with the Camden & Amboy Railroad Company in New Jersey. The New Jersey state legislature granted this firm a virtual transportation monopoly that empowered it to prevent any other transportation alignment from being constructed across the narrow, central waist of New Jersey—the most direct overland route between New York City and Philadelphia—for the period between 1831 and 1869. This exclusivity transformed the combined canal and railroad rights-of-way into extremely important alignments.

Second, vast quantities of Pennsylvania anthracite coal were shipped on the D&R Canal to communities along the waterway, to the major New Jersey cities of Trenton and New Brunswick, and also to New York City. This coal was used to power industries and also provided fuel for domestic use. The coal undoubtedly promoted the industrial development of both Trenton and New Brunswick. Maps, atlases, and photographs document such industries spread along the perimeter of the waterway within these cities. Similarly, the steady flow of Pennsylvania anthracite to New York City, via the canal's direct inland route across New Jersey, contributed to the industrial development of this great metropolitan center. Philadelphia received the bulk of its coal via the Pennsylvania transportation network, but at least a percentage of the city's coal requirements was provided by the combination of the D&R Canal and the Belvidere-Delaware Railroad during the second half of the 19th century. This railroad was associated with the D&R Canal and Camden & Amboy Railroad joint companies. It extended along the New Jersey side of the Delaware River and was designed to connect the Delaware and Lehigh river valleys with Trenton, the railway's southern terminus. Lehigh Valley anthracite coal, the primary commodity shipped on the Belvidere-Delaware, was off-loaded at Trenton and then trans-shipped via the D&R Canal or by railroad to Philadelphia and points east. Furthermore, at least a fair percentage of the capitalists within the Pennsylvania anthracite coal industry were based in Philadelphia and derived the economic benefits of shipping via the D&R Canal's direct inland route to New York City.

Finally, even into the early part of this century, the D&R Canal was still an important link between Philadelphia and New York City. It was utilized by the U.S. Government during World War I as a safe inland passage for shipment between the Philadelphia and Brooklyn Naval Yards. One day, perhaps, a cache of D&R Canal Company corporate records may yet be discovered in the corner of some musty warehouse that may shed light on the full scope of the waterway's impact on the Philadelphia and New York metropolitan areas.

The Canal Dredging Project

The author's introduction to the D&R Canal occurred in 1982 during the planning phases for a dredging project that was to remove silt from 32 miles of the waterway (19 miles of the Feeder Canal and 13 miles of the Main Canal) extending between the Prallsville Lock in the Borough of Stockton, Hunterdon County, and the Kingston Lock in Kingston, South Brunswick Township, Middlesex County, New Jersey. As this ambitious undertaking required the draining of the canal, it provided the first opportunity in at least 50 years to observe and document the canal's structure. Over the next three years, the canal prism and its associated structures were observed and documented with field notes, just over 100 drawings, and many hundreds of photographs.

One important piece of information gathered during this period related to the construction of the canal's liner and the waterway's embankments. Various materials had been used to line the canal — mixtures of sand and clay, sometimes with gravel and cobbles; alternating layers of sand and clay; compact sand and gravel. At first, it was expected that the canal-bed liner would consist primarily of clays, but it was later discovered in John B. Jervis's memoirs that the use of sand and gravel liners in canal construction was not unusual (FitzSimons 1971). In fact, Jervis promoted the use of clay and sand mixed with fine gravels as liner materials since he found them to be less susceptible to leakage than the more typical clay liner. When first applied, the clay and sand liner tended to leak, but the leakage helped to compact the sand into a hard, impervious layer that did not crack in frost or allow leaks to enlarge themselves, as was the case when clay alone was used (FitzSimons 1971: 55-57).

Survey of the canal documented a variety of historical approaches to the construction of liners, approaches that represented adaptations to local conditions and materials. In some places, the natural soils had been excavated and formed an impermeable liner. In other areas, shale had been carved out to form the curved or bowl-shaped canal bottom. Most surprisingly, the built-up supporting canal embankments were shown to consist oftentimes of sandy cores capped with clays. In still other instances the entire manmade embankment consisted of clay.

As already mentioned, both banks along the 42-mile D&R Main Canal were continuously lined with laid-up stone that protected the banks from erosion caused by the wakes of passing boats. Great variation in the design of this embankment stone was noted and documented during the dredging project. Remarkably, most of this dry-laid stone survives, even though lack of maintenance has allowed trees to sprout within and near it. The recreational lobby is strongly in favor of the preservation of these trees which they claim form a "natural" landscape. On the other hand, the historic preservationists call for the removal of the trees since they never would have been allowed to grow along the canal edges during the navigational period, and root action is destructive to the historic embankment and the laid-up stonework. Thus far, the recreational interests have prevailed; however, the NJWSA has recently initiated a program to regularly cut brush and trees in and near the masonry walls of the canal's culverts, locks, waste gates, etc.

The dredging project also provided a chance to examine timber cribbing bridge abutments and bridge approach fenders along the canal. The D&R Canal's bridges had on-grade approaches with no clearance for vessels to pass beneath them. Swing bridges were employed to allow boats to pass. Each swing bridge had an abutment at its pivot end that protruded out into the canal. A system of timber fenders or bumpers was required to guide the boats around these abutments. Evidence of the stone-filled timber crib bridge abutments and fender systems was discovered during the dredging project at many of the bridge crossings and in association with canal basin entrances.

Other historic canal features observed and documented during the dredging project included timber floors and miter-gate sills in locks, plank sheathing on the tops of stone culverts as they passed along the bottom of the canal bed, plank flooring and sheetpiling in association with waste gate structure inlets, and a portion of an old canal dredge.

Rehabilitation of Historic Canal Structures

Once the canal dredging project was completed, the NJWSA began a program to rehabilitate many of the historic structures along the waterway. To date, Historic Conservation and Interpretation, Inc., has participated in well over a dozen of these projects, either as a subconsultant to the NJWSA's engineering consultant, or under contract directly to NJWSA. Due to the D&R Canal's listing on both the National and New Jersey Registers of Historic Places, any undertakings on the canal require review and approval by the Office of New Jersey Heritage, the New Jersey Historic Sites Council (both part of the New Jersey Department of Environmental Protection and Energy), and the Delaware and Raritan Canal Commission.

The active construction phases of some of these rehabilitation projects, which have largely centered around historic masonry culverts and waste-gate structures, were completed within the last two years. Much has been learned about the historic design features of these structures, and sensitive rehabilitation approaches have been developed and improved upon as a result of the field experience. For example, it was observed that the 130-foot masonry canal culverts, whether small (single-barrel) or large (triple-barrel) structures, were typically built on timber grillages that served the double roles of foundation and culvert floor. The timber grillage system was apparently used unless bedrock occurred at the proper elevation for the same purpose.

Vertical plank sheetpiling was typically found in association with the culvert inlet and outlet headwalls and within the canal embankments to prevent leakage and erosion beneath the structure's walls and at the junction of the structure with the canal embankments above it. Finally, a system of plank sheathing was present on top of the culvert's masonry barrels in the bottom of the canal bed. This sheathing was attached to timber nailers embedded in the masonry arches and prevented water leakage through the tops of the culverts. The planking was used in conjunction with a skim coat of mortar on top of the masonry and sand/clay canal liner in the canal bed on top of the plank platform. The wood employed on the culverts has been identified as eastern white pine (*Pinus strobus*).

These culvert construction techniques were quite similar to a specification written by D&R Canal engineer Canvass White for the 1830 construction of the Shabakunk Creek culvert (a short distance north of Trenton) beneath the D&R Canal. A portion of this specification (White 1830-1834), one of only a few known accounts of the construction of original D&R Canal structures, is excerpted here:

> The said culvert N 5 to be constructed with three arches of twelve feet span each - the abutments and piers to be four feet thick and of such height and length as may be directed by the Engineer. The arches to be eighteen inches thick at the commencement on the abutments & piers and gradually diminish to fifteen inches at the top. The wing stone & key stone [*sic*] of the arches and the coping of the parapet [*sic*] walls + wing walls to be cut stone. The coping to be one foot thick and eighteen inches wide and stones not less than three feet in length [Note: White appears to use thickness to mean the height of the exposed face of the stone in the wall, width to mean the depth of the stone extending into the wall's interior, and length to mean the horizontal axis of the stone exposed in the face of the headwall.] and dowelled together with cast iron dowells [*sic*] eight inches long passing four inches into each stone to be one & a half inches square. Every other block of the wing stone shall be double the width of the adjoining stone and none [*sic*] less than twelve inches wide and twelve inches thick, at top or upper edge, the arches to be good flat stone with even edges and not less than six inches thick & twelve inches wide or more than twelve inches thick. The wings to be built such height, form & thickness as shall be directed, the whole to be laid in hydraulic cement, the sand to be washed clean from all dirt or impurity and used in such proportions with the cement as may be directed, and such quantity of clean gravel mixed with a portion of the mortar or cement as may be directed to fill the interstices between the stone and grouted when & where required. [Note: White was the first to discover hydraulic cement in this country; it was first employed during the Erie Canal's construction. Canvass White's brother, Hugh, appears to have been the primary supplier of cement for the D&R Canal's construction.] The foundation to be laid with timber as directed and covered with two inch pine plank well spiked or treenailed. A row of sheetpiling shall be placed the whole length of the foundation along the upper edge not less than four feet in depth and secured to the foundation timber. The top of the culvert to be covered with good seasoned three inch pine plank not exceeding ten inches in width, jointed and laid water tight in a line with the canal, secured by spikes or treenails to

timbers laid and secured in the mason work in a line with the arches. Said floor to extend under the banks of the canal ten feet to a range of sheetpiling or yoke to be well secured to and connected with the stone work the plank of this first sheetpiling to be three feet in height above the floor or top of the culvert, and at ten feet distance to be placed another range of sheetpiling or yoke four feet in height and backed by a wall two feet thick and three feet high well bound into the arches and the stone work between the arches, the said yokes or sheetpiling to extend the whole length of the culvert under each bank of the canal and pass over the abutments and well connected with the foundations and extending back of the walls into the embankments at least six feet. The floor or lining of the culvert to be connected with sheetpiling along the abutments and wings, and an apron so arranged as to join the culvert with the embankments safe and water tight.

During the recent NJWSA culvert rehabilitation projects, the stone headwalls and arches, consisting of native shales and sandstones, were repointed or partially rebuilt, plank interior flooring was repaired or partially replaced, canal embankments and associated embankment stone above the culverts partially or entirely rebuilt, and the top of culverts beneath the canal bed have had masonry and plank sheathing repair followed by the reinstallation of clay liner. In addition, archival-quality photographs were produced for two of the triple-barrel culverts and submitted to the Historic American Engineering Record (HAER) for the Library of Congress collection.

Rehabilitation of two 19th century waste gates, the Kingston Gate and the Griggstown Gate (see Figure 2) also provided much information about the canal's original construction design and techniques. These gates allowed excess water in the canal to be drained, and also provided a means of sending water into lock bypass channels.

The original waste gates on the D&R Canal had an open, stone-wall-lined sluiceway channel that pierced the towpath embankment and was spanned by a timber towpath bridge. The timber gate was located at either the canal edge of the sluiceway or at the end away from the canal. Early in this century, the bridges associated with the open sluiceway style gates were removed, large concrete pipe culverts were installed, and the gap in the embankment was filled to create a continuous on-grade towpath (Figure 3). All of the timber gates were then relocated to the canal end of the concrete pipe culvert. Fortunately, the conversion of the waste gates usually left intact the original 19th century stone inlet and outlet wingwalls and the sluiceway channel walls extending through the towpath embankment.

In the 1950s, the State of New Jersey replaced the timber gates with metal gates, and concrete headwalls were added at the inlet ends of the structures. These new walls partially obscured the original stone inlets but

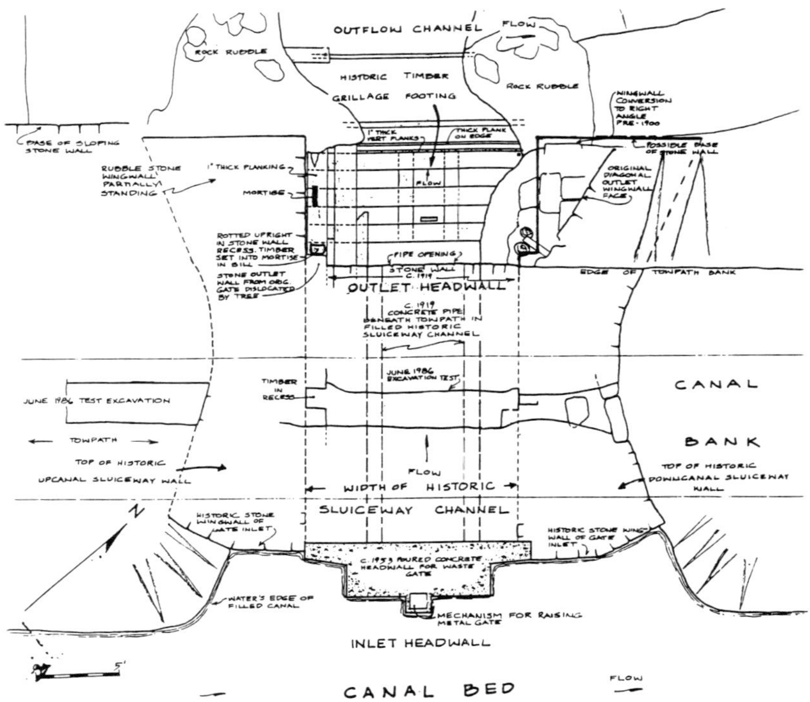

FIGURE 3. Plan of archaeological remains at the Griggstown Waste Gate on the D&R Canal.

left them in place. Following many decades of neglect and unchecked growth of vegetation within and near the old stone masonry, the original waste gates had deteriorated and, in some instances, had partially collapsed. As part of the recent rehabilitation projects, the Kingston and Griggstown waste gate structures were extensively repaired and partially reconstructed.

During this work on the gates, vertical plank sheetpiling was unearthed in the canal bed at the base of the stone inlet headwalls. The planking had been used in conjunction with a sand/clay liner to prevent infiltration of canal water beneath the masonry structure. Plank floors or aprons, also present in the canal bed in front of the gate inlets, were designed to prevent scouring of the canal bottom when water rushed through the open gates.

Beneath the masonry structure of the waste gates' headwalls and the sluiceway walls that extended through the towpath embankment was a timber footing or grillage system consisting of four layers of timbers and planks. The main timbers measured 8 inches x 12 inches x 30 feet long. These timbers rested on flat, 11-inch x 2-inch-thick planks that in turn were supported by 2-inch-thick on-edge planks. The whole grillage system had compacted soils beneath and surrounding it. Short, vertical 2-inch-thick planks attached to the sides of the main timbers were also driven into the compacted ground to further stabilize the grillage system. Finally, a layer of 2-inch-thick planking topped the timbers and created a platform or deck upon which the masonry structure of the waste gate had been erected. Additional rows of 2-inch-thick vertical plank sheetpiling, extending into the ground to a depth of 4 feet, were attached to the outlet edge of the grillage and within the interior of the grillage itself. Once again, this planking prevented infiltration of canal water under the masonry walls and grillage and protected the outlet end of the structure from erosion as canal water rushed through the opened gate. An extensive plank-floor-lined outflow channel guided water away from the gate to a nearby river.

At the Kingston waste gate, one of the stone outlet wingwalls was intact and required only repointing; its companion wingwall, however, was almost completely collapsed and had to be reconstructed. Stone that had fallen from the original wall was salvaged and reassembled to approximate the coursing of the surviving wingwall. The most difficult component to reconstruct was the sloping top of the wall, which had large coping slabs notched to fit together. After detailed examination of these pieces, the "puzzle" was fitted back together, and the coping was re-established as authentically as possible. The timber grillage system survived intact beneath both wingwalls and was reused in the new wall's reconstruction. The same grillage supported a plank outflow apron at the structure, which was also replanked as part of the rehabilitation. Finally, the vertical plank sheetpiling at the outlet edge of the grillage was re-established.

At the Griggstown waste gate the outlet wingwalls were so deteriorated that they had to be dismantled and completely reconstructed (Figure 4). Their condition was so poor that it was not possible to determine original construction details based on the standing ruins. Rather, a combination of archaeological evidence at their bases and an extraordinary 1904 photograph that showed the gate's outlet from a distance were employed to effect a reasonably accurate restoration of the walls.

Following the dismantling of the wingwalls it was possible to examine the gate's timber grillage foundation in detail. Due to a fluctuating groundwater table at the gate outlet, the grillage had not been consistently wet and therefore had deteriorated. As a result, new concrete foundations had to be installed to allow the reconstruction of the two wingwalls. The walls were rebuilt by a skilled mason who was able to recreate the coursing that appeared in the 1904 photograph and to shape the coping stones to form the sloping tops of the walls. The resulting reconstruction of the waste gate's outlet was as historically accurate as possible, based on the available information (Figure

FIGURE 4. View of deteriorated Griggstown Waste Gate outlet in 1989, prior to rehabilitation. *Photo: Brian H. Morrell.*

5). The project also entailed repointing the gate's stone inlet headwall; however, the early 20th century concrete pipe culvert through the towpath was left in place, and no attempt was made to reopen the historic sluiceway channel.

At present the author is serving with Historic Conservation and Interpretation, Inc., as a consultant to the NJWSA on the rehabilitation of five additional historic structures on the D&R Canal — two culverts, two waste gates, and a stone-paved overflow spillway. The last of these masonry structures is 130 feet long and forms an overflow within part of the towpath embankment. For all of these projects, the preliminary planning phases, including the initial documentation and archaeological field work, have been completed. The active construction phase for one of the culverts was completed during the fall and winter of 1991-1992. The actual rehabilitation on the balance of these projects will occur during the next two years and will afford additional opportunities to document D&R Canal construction information.

Rehabilitation Philosophy

Throughout these canal structure rehabilitation projects, an attempt has been made to gather as much information as possible during the initial phases to understand the original designs of the historic structures. Based on these findings, rehabilitation procedures have then been recommended that remain as true to the original designs as possible; however, when the original designs included structural flaws or weaknesses, the rehabilitation seeks to improve upon the old work. A major goal is to preserve and reuse

FIGURE 5. View of rehabilitated Griggstown Waste Gate outlet in 1990. *Photo: Brian H. Morrell.*

as much of the original construction as is possible. This approach is applied to the entire structure, not just to those portions that are normally above ground and visible.

One of the most critical factors in this work, and the most difficult to achieve, is the thorough examination and documentation of the historic structures, a process that enables a fuller understanding of the original construction techniques as well as the source of any existing deficiencies. A typical problem is canal leakage through masonry structures. It is essential to discover the sources and pathways of such leaks before plans are devised to stop the infiltration. The physical examination and analysis of the canal and canal structures themselves is all the more critical since virtually no engineering records are known to exist for the D&R Canal. Most of what is learned about 19th century canal-building technology on this important waterway must come from the study of the physical remains of the canal itself.

The field work needed to document the canal and its associated structures is an enormous task considering that it usually requires draining at least part of the canal adjacent to the structure, perhaps removal of portions of the canal bed, and even excavation of sections of the canal embankments. Such procedures require substantial construction activities and are further complicated by the need to maintain constant flow in the canal due to its present-day function as a public water supply system. Indeed, the situation is even more complex since the planning process for the active rehabilitation phase, including the development of design documents, the receipt of the necessary permits for the project, and the public bidding process, can require about one year's time between the initial exploration of the structure and the beginning of active reconstruction. During this interval, the canal bed and embankments cannot be left vulnerable to damage from flooding and breaching. In addition, there are public safety issues since the canal is also a state park, and there is the need to maintain flow in the canal for water supply purposes.

One great assistance in the exploratory phase has been the use of "Porta Dams," which are fabric-lined dams erected on a steel framework. Such dams are vastly less expensive than conventional clay coffer dams and are easily installed and removed without damaging the historic fabric of the canal.

Concerns about maintenance of water supply, potential damage to the canal, and the expense of thorough examination and documentation oftentimes coincide to limit the amount of work that can be conducted during the initial phases of a project. In such cases, it is vital to design the project with as much flexibility as possible, allowing for some adjustments in the rehabilitation approach following the complete exposure of the historic structure during the active construction phase. It is very difficult to allow sufficient flexibility in construction design without also subjecting the agency undertaking the project to extremely expensive construction change-orders while under a construction contract. However, the work accomplished on the D&R Canal to date has made it possible to predict many of the major preservation issues that may arise. Rehabilitation work on this canal is therefore important knowledge for those undertaking the repair and reconstruction of other historic canals and their associated structures. On the other hand, there are always some surprises when such historic features are completely revealed.

Finally, it is important to emphasize that consulting engineers who provide both the design work for the rehabilitation project and construction management must be knowledgeable about historic engineering structures and should be sympathetic to their preservation. Without such understanding and appreciation for these engineering works, it is all too likely that rehabilitation approaches will be formulated that do not respect the original structure and that will result in unnecessary alteration or removal of its historic fabric.

The multiple uses of the D&R Canal today — as a public water supply system, a recreational facility, and an historic district — present challenges in managing the facility for the public benefit while at the same time preserving its cultural integrity. Nevertheless, such a goal is attainable, and the author continues his efforts to ensure that sympathetic rehabilitation efforts are being employed.

References

FitzSimons, Neal, editor

1971 *The Reminiscences of John B. Jervis Engineer of the Old Croton*. Syracuse, N.Y.: Syracuse University Press.

McKelvey, William J., Jr.

1975 *The Delaware & Raritan Canal, A Pictorial History*. York, Pennsylvania: Canal Press, Inc.

White, Canvass

1830-1834

Personal Papers of Canvass White, Including Papers, Letters, and Specifications for Work on the Delaware & Raritan Canal. Collection 609. Special Collections, Department of Manuscripts and University Archives, Cornell University, Ithaca, New York.

CONCLUSION

Emory L. Kemp

Canals hold a particular fascination not only for legions of amateurs, who have organized themselves into numerous canal societies, but also for professional archaeologists and historians. The subject provides a rich historical record as well as evidence obtained from archaeological field work. From the internalist's point of view, historians of technology can trace the development of the pound lock from the fruitful mind of Leonardo da Vinci in the 16th century to the development of incline planes and other hydraulic devices promoted by Robert Fulton and others at the beginning of the 19th century. From Leonardo's invention of the pound lock with miter gates, it is possible to trace the development of canal technology not only in Europe but in America. As a generic term, "canals" can refer to artificial channels dug out by horse and hand as well as to river navigations that combine locks and dams to create slackwater pools. Such devices as wickets, gates, weirs, and other hydraulic equipment can be traced from nation to nation as well as from canal engineer to engineer. The earlier work of Jessop on the Lehigh Navigation Canal and other works, for example, can clearly be seen in his design for locks and dams on the Delaware and Raritan Canal. In a similar way, the famous incline planes on the Morris Canal are reflected in the later construction of the Shubenacadie Canal in Nova Scotia, Canada.

Canals can preserve a remarkable array of technological themes and variations that enhance the value of archaeological studies. Canal locks, for example, traditionally used miter gates with wickets, but bear-trap gates were used earlier, particularly on the Lehigh Navigation Canal, as a result of an invention by Josiah White and Erskine Hazard. Vertical lift or portcullis gates were employed, particularly in the upper end of locks, but so too were drop gates on a number of canals. On wide canal chambers horizontally sliding gates were also employed, such as those on the Kanawha Navigation Canal. The filling and discharging of canal chambers was greatly improved by the development of side-filling locks and, later, bottom-filling locks which had control valves mounted in the conduits. This alleviated the need for wickets, which were manually operated and mounted, in the gates themselves. Although not common, mules were occasionally replaced with mechanical hauling systems similar to ski lifts. The power-operated cable systems allowed vessels to be moved quickly through locks such as the tidelocks on the Delaware and Raritan Canal, or the well-known electrical mules used on the Panama Canal. Thus, the study of archaeological evidence can greatly enhance the historical record when it comes to understanding the construction and operation of canals and river navigations.

Although one may subscribe to the idea that the artifact is evidence, the artifact does not itself speak or leave a record and indeed may require careful examination to glean any useful information. Archaeologists and other students of technology and culture have moved a long way from Lord Acton's idea that simply organizing evidence in a chronological order will result in a clear historical message. In the case of industrial archaeology, the documentation of a site now consists of the preparation of site plans, measured drawings and archival photography that complements the drawings, as well as the writing of a contextual history of the site. This has become the standard approach to site documentation established under the leadership of the Historic American Engineering Record (HAER), which was established in 1969.

Not all structures or industrial artifacts need such complete documentation. With limited funds, one has to decide clearly what has to be documented and how the evidence is to be interpreted or utilized. For instance, measured drawings on canal structures to be used for illustrative purposes in a publication vary greatly from the detailed information required if one is to restore a canal lock. In order to evaluate the technical significance of a particular canal structure, it is necessary to compare its design to that of other contemporary structures. Unless the investigator has a personal knowledge of a wide number of canals in the country, it is essential to establish a data base which can retrieve significant information on all manner of hydraulic structures. There is an urgent need for such a data base not only for historians and archaeologists but for all those involved in the preservation of engineering works.

Contextual Perspectives

An economic or social historian may view the construction and operation of canals in a broad context showing how canals affected not only the economy of a region, or indeed the nation, but also the influence of the canal on the everyday lives of people. One of the broad historical themes that is used to explain the historical significance of canals and river navigations is the internal improvements movement. To many people this movement began with the publication in 1808 of Albert Gallatin's well-known report on internal improvements (Gallatin 1968 [1808]), in which he indicated that the development of a network of transportation systems of roads and canals, and much later railways, would enable the new Republic to exploit the vast resources of the interior of the country, particularly in those states formed from the Northwest Territory.

The internal improvements movement did not begin with Gallatin but rather predates the formation of the United States. Washington and others in the Old Dominion promoted the development of canals. Washington proposed the development of the Potomac Canal to reach the Ohio River through the Casselman and Youghiogheny rivers. Farther south, he promoted the development of the James River and Kanawha Canal to link the James River, which flowed into the Atlantic, with the Kanawha River, a major tributary of the Ohio River. In other sections of the Northeast, early canals were also promoted before Gallatin's report was issued. His report, then, recommended the involvement of the federal government to develop a rational nationwide system of roads and canals. Gallatin's ideas were never implemented, however, and the federal government played a very limited role in the internal improvements movement except for the building of the National Road. Thus, rather than developing a rational network to serve the new nation, the internal improvements movement became one of intense competition between various ports on the East Coast, each vying for access to the Midwest. Unlike the development of canals in Britain or France, the canal system in the United States was built on the basis of mixed capital. In the case of Virginia, for example, both state-appropriated and municipal funds were provided as were the resources of private investors.

The French were leaders in the development of canals. The Briare Canal dates to the beginning of the 17th century. By 1605, the great seven-staircase lock (sept écluses) on the Briare represented a notable engineering achievement. By the end of the 17th century the famous Canal du Midi was completed under the direction of Riquet, the chief engineer, and Colbert, the famous financier of Louis XIV. It was the Canal du Midi that inspired the Duke of Bridgewater to return to Manchester, England, and construct the first British canal. The building of the Canal du Midi, like the American interstate highway system, was promoted on the basis of defense as it provided an interior access between the Bay of Biscay and the Mediterranean Sea. It was constructed in the grand manner and facilitated not only the movement of military supplies but also was a commercial success. In the British case, canals were built solely with private funds. The role of the government was simply to provide the means for acquiring land and funds for their construction. The mixed capital system of the American canals thus stands midway between the French and British systems of canal financing.

The Duke of Bridgewater's canal was finished in 1759 and provided a very economical method of transporting coal to Manchester, the leading British industrial city. From the opening of the Bridgewater Canal to the 1820s, canals remained the unchallenged transportation system for the movement of heavy raw materials and manufactured goods. The British canals were to become the arteries of the industrial revolution, connecting already established urban industrial areas. In the American case, canals were built to develop the interior of the country from the East Coast, and in many cases they passed through an untamed wilderness that provided little revenue until the entire canal was completed. Whereas the British had more than 60 years to develop a canal network, the American situation was quite different. The

Erie Canal, which was an enormous success and started a canal-building boom, was not completed until 1824, when railways already had begun to emerge as an alternative system. Thus, one must view the American canal-building era in the context of the rapid development of private railway companies. In the beginning, the latter were regarded as complementary to canals by providing tributaries through which goods might be transported to the nearest canal. In many cases the new technology devoured the earlier system. One has only to think of the symbolism of the James River and Kanawha Canal when the Chesapeake and Ohio Railway laid its tracks on the towpath of the earlier canal.

From the technological point of view, the artificial or cut canal had severe limitations. Because of its comparatively narrow width and shallow depth it was not possible to provide steam propulsion on any but a limited number of canals. The wash from stream tugs would heavily erode the canal bank; thus, these man-made waterways were limited to the movement of goods drawn by horses or mules. As such, they simply could not compete with the railways over the long term. The most significant developments were in river navigations beginning with the complete control of rivers like the Great Kanawha with a series of locks and dams. These navigations have been continually upgraded until, today, the standard lock on the Ohio River is 110 feet wide and 1,200 feet long. These river navigations provide a very efficient means of moving raw materials, particularly coal, but also to a large extent petroleum, chemicals, and grains. Many of the present structures completed in the early part of this century are now being heavily modified or replaced. They, too, should be the subjects of historical and archaeological inquiry.

As in the case of most archaeological investigations, the papers in this symposium deal with specific sites such as the Santee Canal in South Carolina or the Delaware and Raritan Canal in New Jersey. These papers give a fascinating insight into the construction and operation of particular canals and the effects that they had on nearby cities and towns. The true significance of these canals, however, can be established only by viewing them in the larger context of the technological, economic, and social history of the country.

References

Gallatin, Albert

1968 *Report of the Secretary of the Treasury* [Albert Gallatin] *on the Subject of Public Roads & Canals*. 1808. Augustus M. Kelley, New York: Reprints of Economic Classics.

Contributors

Scott F. Anfinson is employed by the Minnesota Historical Society as the National Register archaeologist in the Minnesota State Historic Preservation Office. He has worked at the Minnesota Historical Society since 1975. He received his B.A. degree in anthropology from the University of Minnesota in 1972, his M.A. degree from the University of Nebraska in 1977, and his Ph.D. from the University of Minnesota in 1987. His research interests include Plains archaeology, urban archaeology, ceramic and faunal analysis, and paleoecology. He is also the director of Minnesota's Lake Superior shipwrecks initiative. Anfinson has been involved in extensive archaeological studies of the Minnesota waterfront, one of the most important features of which is the waterpower canal that he discusses in this volume. Two recent volumes of *The Minnesota Archaeologist* (Nos. 48 and 49) written by Anfinson discuss the archaeology of the Minnesota waterfront.

Ronald C. Carlisle is Senior Historian and former director of the Cultural Resources Section in the Transportation Planning Group of Michael Baker Jr., Inc., an engineering and environmental services firm in Pittsburgh, Pennsylvania. He received his B.A. in anthropology and history and his M.A. and Ph.D. degrees in anthropology from the University of Pittsburgh, where he was research assistant professor of anthropology from 1985 to 1991. From October 1989 to May 1991, Carlisle was the interim director of the Cultural Resource Management Program at Pitt. He is currently adjunct associate professor in the department of History at that university. From 1977 through 1985, Carlisle was the associate editor of *Historical Archaeology*, and a member of its editorial advisory committee. He is a former member of the Historic Preservation Board of Pennsylvania and a former research associate of the Division of Anthropology, The Carnegie Museum of Natural History, Pittsburgh, Pennsylvania. Carlisle is a member of the Society for Historical Archaeology's Urban Archaeology Group and the Committee on Pittsburgh Archaeology and History. His research interests include urban archaeology, African-American archaeology, vernacular architecture, and the study of military fortifications. He is the compiler and editor of *Americans Before Columbus: Ice-Age Origins*, selected papers from the Smithsonian Institution's first Columbian Quincentenary Symposium, published in the *Ethnology Monographs* series by the Department of Anthropology, University of Pittsburgh, in 1988.

Verna L. Cowin is associate curator of anthropology at the Carnegie Museum of Natural History, Pittsburgh, Pennsylvania. She is a graduate of Moravian College and received both M.A. and Ph.D. degrees in anthropology from the University of Pittsburgh. In 1982, she conducted archaeological excavations at the construction site of the PPG building in Pittsburgh, work that later formed the subject of an archaeological exhibition at the Carnegie Museum. Her 1985 dissertation, *Pittsburgh Archaeological Resources and National Register Survey*, serves as a guide for resource management in downtown Pittsburgh and for the city's North Side and South Side neighborhoods. In 1987, Cowin served on an *ad hoc* committee to negotiate the documentation and preservation of the Pennsylvania Canal lift and weigh locks in Pittsburgh that she and her co-authors discuss in this volume. She continues to contribute to plans for the reconstruction of the lift lock and for its public interpretation. Cowin is a member of the Society for Historical Archaeology's Urban Archaeology Group and the Committee on Pittsburgh Archaeology and History. She is also a member of the Historic Preservation Board of Pennsylvania.

Jack B. Irion is vice-president of nautical archeological services for R. Christopher Goodwin & Associates, Inc., a cultural resources consulting firm in New Orleans, Louisiana. Prior to joining this firm in 1991, he was the manager of the cultural resource section of GAI Consultants, Inc., a Pittsburgh-based engineering firm. In 1987 he directed the documentation and archaeological excavation of the lift and weigh locks that he and his co-authors discuss here. Irion received M.A. and Ph.D. degrees from the University of Texas, Austin, and specializes in nautical archaeology. He has directed underwater archaeological projects throughout the United States, Mexico, and Belize. Much of his research focuses on 19th-century watercraft, vessels of the Civil War, and Confederate coastal fortifications. Irion's earlier work on riverine and marine craft afforded him a unique perspective on the excavation of the Pennsylvania Canal features in Pittsburgh.

Emory L. Kemp worked for leading engineering and consulting firms in England before receiving his Ph.D. in theoretical and applied mechanics from the University of Illinois. He joined West Virginia University in 1962 to establish a graduate program in structural engineering. Later, he founded WVU's program in the history of science and technology. Fostering the use of the material culture approach for the study of the industrial past, he has researched and preserved historic industrial sites around the country and has advocated their public interpretation. Kemp was a founding member, and recently president, of the Society for Industrial Archaeology, and he has been a member of the advisory board of the Historic American Engineering Record. As a Regent's Fellow at the Smithsonian Institution in 1983-1984, he researched the history of suspension bridges for a forthcoming book. He specializes in the study of historic bridges and has been involved in more than 36 documentation and restoration studies of bridges and other industrial structures. Emory L. Kemp is currently director of the newly established Institute for the History of Technology and Industrial Archaeology in Morgantown, West Virginia.

Lance E. Metz is historian of the Canal Museum at the Hugh Moore Historical Park and Museums, in Easton, Pennsylvania. He received his B.A. *magna cum laude*, with Honors in History, from Moravian College in 1970. He received an M.A. in Maritime History from the University of Maine at Orono in 1973. He has also undertaken additional graduate work at Lehigh University. Metz is one of the country's leading experts on the history of canals and canal technology. He is the editor of the *Proceedings* published by the Canal History and Technology Press and is the co-author with Ann Bartholomew of a 1989 publication on the Delaware and Lehigh Canals, published by the Center for Canal History and Technology.

Brian H. Morrell has been affiliated with the cultural resource management firm Historic Conservation & Interpretation, Inc., (HCI) of Newton, New Jersey, since 1973. He graduated *magna cum laude* from Fairleigh Dickinson University in New Jersey with a B.A. degree in anthropology. He specializes in industrial archaeology, historic documentation, and historic preservation projects with HCI. Morrell has participated in and supervised dozens of archaeological and rehabilitation projects on both of New Jersey's 19th-century canals, the Morris Canal in the northern part of the state, and the Delaware & Raritan Canal in central New Jersey. In the process, he has become well versed in historic canal construction techniques and has documented numerous masonry canal structures, advising on their rehabilitation. As an independent consultant, Morrell conducted an archaeological and historic preservation survey in 1983 along a 33-mile length of the Morris Canal in northwestern New Jersey for the Warren County Planning Board. His survey results were published by the Morris Canal Committee and the Warren County Board of Chosen Freeholders in 1987 under the title *Historic Preservation Survey of the Morris Canal in Warren County, New Jersey*.

Mark M. Newell is a staff member of the division of underwater archaeology at South Carolina Institute of Archaeology and Anthropology, where he is head of compliance and review activities for federal and state project permitting in South Carolina. These activities include state, federal, and commercial-sector liaison in Columbia, South Carolina, and the work of an archaeological survey team in Charleston, South Carolina. Newell holds an M.A. degree in journalism from the University of London, and he is currently completing a doctoral degree at St. Andrews University's Scottish Institute of Maritime Studies. He is a member of the Royal Archaeological Institute, London; a board member of the Archaeological Research Trust, University of South Carolina; and chairman of the archaeological committee of the American Canal Society. Newell's research interests are in the transfer of industrial and agricultural technologies from the Old World to the New, origins of American maritime technology, and technology of canals and rice-culture hydraulics. Based upon his excavation of a one-mile section of the Old Santee Canal near Charleston, South Carolina, a $3,000,000 canal museum has recently opened at Monck's Corner, South Carolina. Newell is currently consulting on the master plan for the Old Savannah-Ogeechee and Augusta canals in Georgia. His most recent publications on southern canal technology include a two-volume report on the Santee Canal excavations and a publication on that work in progress with the University of South Carolina Press.

Steven J. Shephard has been the assistant city archaeologist with the City of Alexandria, Virginia, since 1980. He received his B.A. degree in history from the University of Florida, a M.A. degree in anthropology from Florida State University, and his Ph.D. in anthropology from Southern Illinois University, Carbondale. Other areas of professional interest include preservation planning and management, socioeconomic stratification, public archaeology, and urban processes. He was closely involved in the excavation and restoration of the tidal lock and basin of the Alexandria Canal detailed in this volume. Shephard has contributed to three books, *Spanish St. Augustine, the Archaeology of a Colonial Creole Community*, by Kathleen Deagan (1983); *Archaeology of Urban America, a Search for Pattern and Process*, edited by Roy S. Dickens, Jr. (1982); and *Consumer Choice in Historical Archaeology*, edited by Suzanne M. Spencer-Wood (1987). Steven J. Shephard is currently the editor of the *Alexandria Archaeology Publications* series.

A. Howe Todd is now retired after 40 years with the City of Richmond, Virginia, where he began to work in the Department of City Planning and where for 15 years he was the assistant city manager. He received his B.S. degree in civil engineering from Lehigh University and his Master of City Planning degree from the Massachusetts Institute of Technology. After retiring from local government in 1986, Todd created a citizens' committee under the auspices of the Historic Richmond Foundation to reconstruct the portion of the James River and Kanawha Canal that passes through Richmond. His committee published a feasibility report on the project, entitled *The Richmond Canals*. This was followed by *The Richmond Canals—Preliminary Design, Eastern End* and another volume entitled *History Along the Waterfront, 1607-1865*. In pursuit of their goal to preserve, restore, and interpret the Richmond Canal, Todd and his committee have raised over $500,000 to date.

William E. Trout, III, is a founder and the current president of the American Canal Society. He attended both the University of Richmond and Indiana University, where he received his Ph.D. in genetics under Nobel laureate H.J. Muller. Throughout his career as a research geneticist, he maintained an active interest in the history of canals, particularly those in Virginia. In 1983, he retired to devote more time to canal research, field work, and preservation. He is the author of a series of regional canal inventories, the *American Canal Guide*, and a series of historic river and canal atlases for Virginia. He is a board member of the Archaeological Society of Virginia and the Virginia Canals and Navigations Society.

Scott F. Anfinson
Ft. Snelling History Center
St. Paul, Minnesota 55101

Ronald C. Carlisle
Michael Baker Jr., Inc.
501 Parkway View Drive
Pittsburgh, Pennsylvania 15205

Verna L. Cowin
Division of Anthropology
The Carnegie Museum of Natural History
5800 Baum Blvd.
Pittsburgh, Pennsylvania 15206

Jack B. Irion
R. Christopher Goodwin Associates, Inc.
5824 Plauche Street
New Orleans, Louisiana 70123

Emory L. Kemp
Institute for the History of Technology and Industrial Archaeology,
University of West Virginia,
Morgantown, West Virginia 26505

Lance E. Metz
Canal Museum
Hugh Moore Historical Park and Museums
200 South Delaware Drive
P.O. Box 877
Easton, Pennsylvania 18044

Brian H. Morrell
Historic Conservation & Interpretation, Inc.
Box 120, R.D. 3
Newton, New Jersey 07860

Mark M. Newell
South Carolina Institute of Archaeology and Anthropology
University of South Carolina
1321 Pendleton Street
Columbia, South Carolina 29208

Steven J. Shephard
Alexandria Archaeology
105 North Union Street
Alexandria, Virginia 22314

A. Howe Todd
Rte. 2, Box 440
Hanover, Virginia 23069

W.E. Trout, III
American Canal Society
35 Towana Road
Richmond, Virginia 23226

Publications

Books

Josiah White, Quaker Entrepreneur, by Norris Hansell. Combination of White's personal papers and correspondence with narrative by Hansell.
ISBN 0-930973-12-7 — $19.95 (hardback)

The Story Of The General Crushed Stone Company, by W. Julian Parton.
ISBN 0-930973-13-5 — $11.95 (hardback)

A Canal Bibliography, compiled by A.G. Zimmerman. A bibliography of canal sources with primary emphasis on canals in the United States and Canada..
ISBN 0-930973-11-9 — $24.95 (paperback)

The Central Railroad Of New Jersey's First 100 Years, by Elaine Anderson. A 238 page history with 185 rare photographs and illustrations.
ISBN 0-930973-00-3 — $14.00 (paperback)

Anthracite And Slackwater, by F. Charles Petrillo. History of Pennsylvania's North Branch Canal, with maps and illustrations.
ISBN 0-930973-04-6 — $16.25 (paperback); ISBN 0-930973-03-8 — $28.75 (hardback)

The Old Middlesex Canal, by Mary S. Clark. A history of the Middlesex Canal.
ISBN 0-930973-06-2 — $6.00 (paperback); ISBN 0-930973-05-4 — $14.00 (hardback)

Sherman's Guide To Hugh Moore Park, by Lance E. Metz. A history of the transportation, industry and surroundings of present-day Hugh Moore Park, Easton, Pa. with map and photographs.
ISBN 0-930973-07-0 — $3.00 (paperback)

The Anthracite Iron Industry Of The Lehigh Valley, by Craig Bartholomew & Lance E. Metz. A definitive history of these 19th Century iron producers.
ISBN 0-930973-08-9 — $17.95 (paperback)

Delaware And Lehigh Canals, by Ann Bartholomew & Lance E. Metz. A pictorial history of the Delaware Canal and Lehigh Navigation from Bristol to Ashley, Pa.
ISBN 0-930973-09-7 — $28.00 (hardback)

Taking Pictures, by William R. Zwikl. A compilation of 164 photos of events, celebrities and scenery celebrates the 47-year career of an award-winning Lehigh Valley, Pennsylvania, photographer.
ISBN 0-930973-10-0 — $9.95 (hardback)

Colonel James Neilson, by Robert T. Thompson. The activities of a New Jersey businessman and promoter of canal, turnpike and railroad transportation.
ISBN None, Rutgers University Press, 1940 — $12.95 (hardback)

Charles M. Schwab, by E.G. Grace. An intimate biography of a steel magnate, privately published.
ISBN None, Bethlehem Steel Corporation, 1947 — $6.50 (hardback)

Canal History And Technology Proceedings

Softbound volumes of papers dealing with the social and political history and engineering of canals, gravity railroads, anthracite coal, anthracite iron, early railroads, etc.
Volumes 1-7—$11.00; Volumes 8-10 & 12-13—$15.00; Volume 11—$18.00 (paperback)
ISSN 0892-3515; Complete set of all volumes (10% off)

Orders should be accompanied by $4.50 for postage and handling for the first book and $.50 for each additional book. PA residents should add 6% sales tax. Send orders and make checks payable to Canal History and Technology Press, PO Box 877, Easton, PA 18044-0877